D1414368

s are available at special discounts for bulk purchases in the United
ns, institutions, and other organizations. For more information, please
Markets Department at The Perseus Books Group, 11 Cambridge
MA 02142, or call (617) 252-5298.

in the United States of America by Westview Press, 5500 Central
Colorado 80301–2877, and in the United Kingdom by Westview Press,
oad, Cumnor Hill, Oxford OX2 9JJ

orld Wide Web at www.westviewpress.com

ecord for this book is available upon request from Library of Congress.
576-7 (hc); 0-8133-6560-0 (pbk)

in this publication meets the requirements of the American National
rmanence of Paper for Printed Library Materials Z39.48–1984.

7   6   5   4   3   2   1

GE

PHILOSOP

W

Copyright © 2001 by

Westview Press book
States by corporation
contact the Special
Center, Cambridge

Published in 2001
Avenue, Boulder,
12 Hid's Copse R

Find us on the W

A CIP Catalog
ISBN 0-8133-6

The paper used
Standard for P

Westview
PRESS

A Member of the Pe

10   9

For Yee-Wah

CONTENTS

ACKNOWLEDGMENTS

Some of the material in this book was presented to philosophy departments at the State University of New York at Buffalo and the University of Western Ontario, as well as to audiences at the Faculty of Medicine of the University of Calgary, the Joint Centre for Bioethics of the University of Toronto, the annual conference of the Society for Applied Philosophy in Manchester, U.K., a bioethics conference at the University of Tennessee at Knoxville, and meetings of the Canadian Bioethics Society in Edmonton and Quebec City. I am grateful to all of these audiences for their constructive criticisms and suggestions. I taught a graduate seminar on this topic in the winter term of 2000 at McGill University and thank the students for discussing the philosophical aspects of human genetics at such a high level. The first draft of the manuscript was written while I was a fellow at the Institute for Ethics of the American Medical Association in 1998–99. I thank the Institute for the fellowship support I received and especially Linda Emanuel and Jessica Berg for protecting my research and writing time while I was at the AMA.

The two anonymous readers commissioned by Westview Press, as well as Jan Heller, gave me very helpful comments on an earlier draft of the entire manuscript. Lainie Friedman Ross gave me extremely constructive criticisms on drafts of chapters 2 and 3. I also benefited from the comments of Jeff McMahan, John Harris, and two anonymous readers for the *Journal of Medicine and Philosophy* on different parts of chapter 5. Some sections of the book's chapters have appeared as articles in journals. These include "Genes, Embryos, and Future People," *Bioethics* 12 (1998): 211–25; "The Ethics of Human Cloning," *Public Affairs Quarterly* 12 (1998): 63–79; "Identity, Prudential Concern, and Extended Lives," *Bioethics* (forthcoming 2001); and "Extending the Human Life Span," *Journal of Medicine and Philosophy* (forthcoming 2001). A section of chapter 3 appears as "Intervening for Medical Reasons," in Brenda Almond and Michael Parker (eds.), *Ethical Issues in the New Genetics: Are Genes Us?* (Aldershot: Ashgate, 2002). I thank the publishers of these journals and Ashgate Publishing Ltd. for permission to use this material. Sarah Warner, my editor at Westview, was very supportive of the project from the time I first submitted the book proposal. Her patience and wise guidance are greatly appreciated. Finally, I am grateful to my wife, Teresa Yee-Wah Yu, for her inspiration and for gracing my life in so many beautiful ways.

—*Walter Glannon*

INTRODUCTION

Recent advances in human genetics have given us the ability to intervene in the process of human biological development controlled or influenced by genes that extends from zygotes and embryos to persons. One type of intervention is genetic diagnosis of embryos created through in vitro fertilization (IVF), before they are transferred to and implant in the uterine wall. These can be tested for genetic mutations that cause or make us susceptible to disease. This gives prospective parents the choice between allowing an embryo to develop into a person with the risk of having a severe disease or disability, or selectively terminating any further development of that embryo. Alternatively, soon it may become possible to prevent, control, or even cure diseases by delivering normal copies of genes coding for critical proteins into cells so that they will function properly. It even may become possible to manipulate genes in order to enhance people's already normally functioning physical, cognitive, and emotional traits. All of these forms of genetic intervention give us considerable control over how many people will exist, when they will come into existence, their identities as persons, and the length and quality of their lives. In this regard, questions about human genetics are inextricably linked to "genesis" questions in philosophy.

The Human Genome Project was created in 1989 by the National Research Council of the National Academy of Sciences and the Congressional Office of Technology Assessment. Its goal was to map and sequence the entire human genome. To oversee this project, the National Institutes of Health established the National Center for Human Genome Research. Its mandate included the use of a small percentage of the Human Genome Project's funds for assessment of and education on the ethical, legal, and social implications of human genetic research. While the ELSI project encompasses three areas of concern, legal and policy issues have strongly influenced the formulation and discussion of the ethical issues pertaining to human genetics. These issues hinge largely on the use of genetic information acquired through testing and screening programs, and they include privacy, confidentiality, unlawful disclosure, and consent. A major rationale for formulating public policy guidelines and enacting legislation in these areas is to protect individuals from discrimination by prospective insurers or employers on the basis of genetic information. For example, one who has tested positive for a disease-causing gene and whose medical record is publicly accessible may find it difficult or even impossible to obtain life and health insurance. Because insurance operates on the idea of pooling risk, and the genetically better off may not want to pool their risk of disease with the genetically worse off, the latter may be rejected by insurance companies. Furthermore, if genetic information were required when applying for work, then employers who pay the health care costs of their employees might refuse to consider one who has tested positive for a gene that only predisposes one to disease.[1]

Although the legal and policy implications of genetics are of enormous importance, my concern in this book will not be with them. Nor will I address the issue of behavioral genetics. For the question of whether or to what extent genes influence or determine how humans act is so complex and controversial that it deserves to be discussed separately.[2] Instead, I will focus on a particular set of philosophical questions generated by genetics. I believe that these questions are more intriguing and weightier than legal or policy questions, in the sense that they can help us to achieve a deeper understanding of our nature as human beings and persons, as well as of our moral obligations to future generations. More precisely, the book explores two related general philosophical questions pertinent to genetics, one metaphysical, the other moral: (1) How do genes, and different forms of genetic intervention (gene therapy, genetic enhancement, genetic diagnosis of preimplantation embryos, and so on) affect the identities of the people who already exist and those we bring into existence? (2) How do these interventions benefit or harm the people we bring into existence in the near future and those who will exist in the distant future? It is worth raising more specific questions germane to (2). Are we morally obligated to prevent the existence of people who would be severely diseased and disabled? Do the claims of people who will exist in the distant future have less moral weight than the claims of those who will exist in the near future? Would genetic enhancement of cognitive, physical, and emotional capacities to raise them above the normal level of functioning for people be morally objectionable and, if so, on what grounds? How could we ensure equal and fair access to both therapeutic and enhancement genetic technologies for all people?

Methodologically, in each of the five chapters I first lay out the relevant biological or medical facts and possibilities involving genetics and then explore the metaphysical and moral implications of them, especially as they pertain to personal identity and our obligations to future generations. I have tried to make the discussion biologically and medically informed to give more substance to the philosophical claims and arguments. Philosophers frequently ignore or oversimplify science in their discussions of metaphysical and moral issues, often engaging in hypothetical thought-experiments that are overly speculative and lacking in plausibility outside of their own field. It should be noted, however, that my main concern is not with biology and medicine. Rather, I refer to these areas only as a necessary (but not sufficient) basis on which to intelligibly discuss the philosophical issues. Moreover, some of the ideas I explore involve what presently may seem to be remote biological and medical possibilities. These include genetic enhancement, human cloning, and extending the human life span on a broad scale. Nevertheless, it is instructive to consider these possibilities in order to test and refine our metaphysical and moral intuitions. It is particularly important to do this because all too often our moral attitudes and laws lag behind biotechnology. And rather than being forced to react to biotechnological findings, we can more effectively address the philosophical aspects of future findings when they occur by anticipating them in advance. Not too long ago, cryopreservation of embryos, cloning of animals, xenotransplantation, and gene therapy to treat human diseases were unthinkable to many people. Although now these procedures are actually practiced, we still have not sorted out all of the moral issues germane to them.

In thinking philosophically about genesis questions, I have been influenced primarily by the work of philosophers Derek Parfit and David Heyd, especially with respect to the nature and extent of our obligations to future generations.[3] The work of Philip Kitcher and Jeff McMahan has influenced my discussion of other philosophical questions generated by human genetics.[4] My presentation of the biological and medical issues has been shaped by the theory of evolutionary medicine, especially as explained in the work of George C. Williams and Randolph Nesse.[5]

While I was working on the final draft of this book, a much-anticipated book by philosophers Allen Buchanan, Dan Brock, Norman Daniels, and Daniel Wikler appeared.[6] Their discussion of the social, political, and policy implications of genetics is much more comprehensive than mine. However, my book differs from theirs in at least three important respects. First, I offer a more comprehensive discussion of genesis questions, of the extent to which genetics and genetic interventions influence personal identity and the quality of future people's lives. This is significant because questions about whether people are benefited or harmed cannot be addressed adequately without identifying *who* they are. Second, the four philosophers just mentioned do not explore the implications of genetically manipulating the mechanisms of aging to extend the human life span. Third, while their account is biologically and medically well-informed, unlike them I discuss biology within an evolutionary framework. This is a function of the conviction that genes can be understood only from the perspective of evolutionary biology.

Rather than draw a fine-grained distinction between "ethics" and "morality," many philosophers and bioethicists use these two terms interchangeably as equivalent normative notions indicating what ought to or ought not be done. Bernard Williams points out that the difference between the terms derives from distinct Greek and Latin origins. "One difference is that the Latin term from which "moral" comes emphasizes rather more the sense of social expectation, while the Greek favors that of individual character."[7] Similarly, for philosopher and legal theorist Ronald Dworkin, "ethics . . . includes convictions about which kinds of lives are good or bad for persons to lead, and morality includes principles about how a person should treat other people."[8] Because of its social content, "morality" better captures the normative aspects of genetic technology, and accordingly I will use this term instead of "ethics" throughout the book.

At a lower level of argument, I do not adopt a single moral theory to address the different issues in the chapters but instead use different theories to support different moral conclusions. My reason for doing this is to avoid a "top-down" method where I employ one theory from the outset and simply apply it to every moral aspect of genetics covered in the book. This would run the risk of artificially forcing the issues to fit the theory. More importantly, it would ignore the fact that different questions have been raised about different moral implications of human genetics and accordingly require moral analysis that is sensitive to these differences. For example, my discussion of the impersonal comparative principle in chapter 2, regarding the quality of the lives we bring into existence and minimizing the amount of pain and suffering in the world, is motivated by consequentialism. In contrast, in chapter 4 the principle that human clones must not be treated solely as means but also as ends-in-themselves is motivated by Kantianism, or deontology. Similarly, my argument in chapter 3 against genetic manipulation of people's cognitive capacities to improve their health status is grounded in deontological reasoning. The two theories in question are not entirely antithetical. Contemporary nonconsequentialists in the Kantian tradition hold that in some cases we are permitted not to maximize overall best consequences, and that in other cases there are constraints on promoting these consequences.[9] This implies that consequences do matter morally and can be promoted to a certain extent. At a higher level of argument, my general concern in all of the chapters is with how people can benefit from or be harmed by different forms of genetic intervention. In this regard, my general theoretical orientation is a consequentialist one.

The structure of the book is as follows. In chapter 1, I lay out the basic features of human genetics, explain the role that genes play in disease, and discuss how these biological features bear on the identities of persons and our moral obligations to the people we bring into existence. Chapter 2 focuses more closely on moral issues, specifically what the genetic information acquired through presymptomatic genetic testing of adolescents and adults, as well as preimplantation genetic diagnosis of embryos, implies for our obligations to existing and future people. I consider whether there are diseases that make people's lives on balance not worth living, and whether this entails a moral requirement to prevent these lives and the existence of the people who would have them by terminating genetically defective embryos. Early- and late-onset genetic diseases are considered in this regard. Also pertinent here is whether people who know that they are at risk of having a late-onset genetic disease as a result of presymptomatic testing have an obligation to share this information with family members who also may be at risk.

In chapter 3, I examine two other forms of genetic intervention—gene therapy to treat diseases, and genetic enhancement of cognitive, physical, and emotional traits and capacities. I analyze these procedures in terms of benefit and harm, permissibility and impermissibility, and equality and fairness regarding access to them. Because the effects of germ-line genetic manipulation will be passed on to offspring, and because it is not known what these effects might be in the long term, generally there are sound biological and moral reasons against this type of intervention. In somatic-cell gene therapy, on the other hand, the effects remain with the individual who is treated. Although it is medically and morally desirable, somatic-cell gene therapy has been effective in treating only a few diseases and has resulted in some deaths. Insofar as all forms of genetic intervention are designed to prevent, control, or cure disease and disability, the general failure of somatic-cell gene therapy provides an even stronger case for testing and selectively terminating embryos with genetic anomalies causing severe disease and disability. With these same medical and moral goals in mind, I give five reasons against genetic enhancement.

Chapter 4 is an analysis of cloning human beings and human body parts. I maintain that there are few compelling biological reasons for cloning full-fledged humans, and that cloning body parts would sidestep many of the moral qualms generated by the idea of human cloning. Nevertheless, I argue that concerns about the loss of personal autonomy and dignity through human cloning are unfounded. These basic human values may be threatened by the purpose for which cloning is used, but not by the procedure itself, which does not necessarily imply the violation of our humanity or personhood. There are two main moral concerns about cloning that involve potential harm to people. First, DNA damage in the donor cell nucleus could result in premature aging and disease in cloned people. Second, as an asexual form of reproduction, cloning practiced on a broad scale could adversely affect genetic diversity and consequently the ability of future people to adapt to and survive in changing physical environments. Because of this, and because genetic material is transferred from the donor cell nucleus to the oocyte, there are unique genetic features of cloning that distinguish it from other forms of assisted reproduction. In response to the claim by some that widespread human cloning is a fanciful idea, I suggest that many parents might be inclined to resort to cloning to prevent disease in their children. I then point out the adverse impact this practice could have on future people.

In chapter 5, I explore the prospect of extending the human life span through genetic manipulation of the mechanisms of aging in general and telomeres and embryonic stem cells in particular. I explore whether this procedure might increase the number of deleterious mutations in humans and how this might affect the ability of people in the distant future to have a reasonably long and disease-free life span. Moreover, I consider what increased longevity would mean for our prudential concern about our future selves, as well as its collective effects on population and the quality of people's lives.

The general aim of this book is to move philosophers, bioethicists, biologists, and readers in general to reflect on the philosophical questions generated by different aspects of human genetics. These questions include the extent to which genes and different forms of genetic intervention influence whether we are healthy or diseased, our identities as persons, the quality of our lives, and our moral obligations to people in future generations.

I

THE REACH OF GENES: BIOLOGY, METAPHYSICS, MORALITY

Genes are stretches of nucleotide base pairs along strands of the DNA molecule, the major component of each chromosome. They specify the amino acid sequences of structural and enzymatic proteins controlling the functions of simple and multicellular organisms from bacteria such as *E. coli* to humans. Genes are the basic units of heredity through which the biological characteristics that define us as humans are transmitted from one generation to the next.[1] While all human organisms share the same basic genetic code, variations among different alleles, or forms of genes, distinguish each of us from other individuals in our species by producing differences in our phenotype, the set of our observable traits. These genotypic polymorphisms play a major role in shaping the unique biological identity of each human organism.

But while genes influence our biological functions in many crucial respects, they do not by themselves completely determine our biological fate. For how cells function at the molecular level depends to a great extent on how the genes in these cells interact with certain chemical processes inside our bodies and brains, as well as with various environmental factors. These include not only ionizing radiation, viruses, and bacteria, but also diet and manipulation of genes through different forms of biotechnology.[2] Furthermore, while genes influence the structures and functions of our bodies and brains necessary to generate and sustain the consciousness and mental life that make us persons, genes cannot offer us a complete explanation of the psychological concepts of personhood and personal identity through time. Nor can genetics tell us what is morally permissible or impermissible, or what constitutes benefit and harm, with respect to the manipulation of genes in the somatic or germ cells to prevent or enhance the expression of certain physical and mental traits. These questions cannot be raised and analyzed within the domain of genetics alone.

Nevertheless, an examination of the subtle and complex ways in which genes causally influence the nature of human biology and personal identity, as well as how they can benefit and harm us, can elucidate such questions as, What does it mean to be human? What is a "person"? Are we essentially human organisms or persons? What obligations do we have to prevent, treat, or cure diseases in people by intervening at the molecular genetic level? How many and what sort of people should we bring into existence in the near and distant future? My aim in this chapter, and more generally throughout this book, is to address these questions by exploring the influence of genes along biological, metaphysical, and moral dimensions.

Although they are interrelated in important respects, each of these three dimensions is necessary but not sufficient for a satisfactory account of the other two. An understanding of the biological features of our bodies and brains is necessary to explain how consciousness and other mental states identified with personhood and personal identity through time can emerge and be sustained. But biological features alone cannot account for the contents of these mental states or their phenomenological, qualitative character. Similarly, an understanding of the metaphysics of personal identity is necessary to account for the moral notions of benefit and harm. For only beings with interests can be harmed, and having interests presupposes the capacity for consciousness and other forms of mental life that define persons. Still, this metaphysical notion by itself cannot fully explain the moral significance of benefits and harms, or why benefits to some people can outweigh, or be outweighed by, harms to others. I will consider these three dimensions in turn, showing how each influences the others, with further elaboration and refinement of the nature of their interrelations in subsequent chapters.

The geneticist Theodosius Dobzhansky claimed that "nothing in biology makes sense except in the light of evolution."[3] In adopting this position, I will maintain that questions about genes and their implications for the metaphysical and moral issues I have mentioned should be raised and analyzed within a framework of evolutionary biology. Our bodies, immune systems, brains, and mental life have developed over a long period of evolutionary time and are integrated parts of a design naturally selected to confer a survival advantage on human organisms in different physical environments. We have been designed as organisms through natural selection to survive in order to enhance our reproductive fitness so that our genes can be efficiently and effectively transmitted into the next generation. Biological fitness is necessary for reproductive success. Unfortunately, for persons with the desire to have longer lives, the protection that genes afford our organisms from acute diseases earlier in life makes our organisms more vulnerable to chronic diseases later in life. I will pay closer attention to this trade-off and what it implies for extending the human life span through genetic manipulation in chapter 5.

It is one thing to explain the impersonal nature of genes in terms of evolutionary biology. It is quite another to show how genes can affect the identities and interests of persons. Genes should be understood as operating at two distinct but related levels. On a general level, genes function according to principles of natural selection, enhancing human survival and reproductive fitness so that they can be passed on from one generation to the next. For humans and other species, this process takes place over many generations and involves extremely long periods of time. On a particular level, genes affect human organisms and persons over shorter periods of time by controlling the synthesis of proteins that underlies the cellular functions of our bodies and brains. Insofar as our mental life causally depends on the structures and functions of our bodies and brains, and genes influence these structures and functions, genes indirectly shape our understanding of ourselves as self-conscious individuals persisting through time. Moreover, by controlling cellular mechanisms genes play a causal role in the etiology of healthy and diseased conditions, depending on whether the alleles of these genes are normal or mutated and the extent to which mutations affect these mechanisms. Because persons have an interest in being healthy and avoiding disease, because benefit and harm are defined in terms of the satisfaction and defeat of interests, and because genes play a causal role in health and disease, genes can indirectly benefit or harm persons. So, while the health or disease that individual persons experience are not directly caused by the way genes operate at the evolutionary level, they are the indirect effects of the force of natural selection.

In the remainder of this chapter, I first discuss the connection between genes and disease. Then I go on to distinguish the biological type "human organism," or "human being," from the psychological type "person" and argue that we are essentially persons. I explain that these two ontological types correspond to distinct biological and psychological senses of "life" and that each of these types has distinct criteria of identity. In addition, I explain how genes and various forms of genetic intervention can affect the identities of persons. Finally, I sketch some of the moral issues for discussion in subsequent chapters pertaining to these interventions, addressing the more general question of what our obligations are to the people we bring into existence in the near and distant future.

Genes and Disease

The Human Genome Project was created in 1989 on the joint recommendation of the National Research Council of the National Academy of Sciences and the Congressional Office of Technology Assessment.[4] Its aim was to decipher all of the roughly 30,000 to 40,000 genes encoded in the 3 billion nucleotide base pairs in our DNA. In 1991, the then director of the project, and earlier codiscoverer of the DNA molecule, James Watson, issued the following statement:

> A more important set of instruction books will never be found by human beings. When finally interpreted, the genetic messages encoded within our DNA molecules will provide the ultimate answers to the chemical under-pinnings of human existence. They will not only help us understand how we function as healthy human beings but will also explain, at the chemical level, the role of genetic factors in a multitude of diseases—such as cancer, Alzheimer's disease and schizophrenia—that diminish the individual lives of so many millions of people.[5]

Clearly, deciphering the genetic code is motivated mainly by therapeutic concerns, and a better understanding of how genes function could lead to better means of preventing, treating, or even curing diseases and promoting health. This reasoning is based on the theory that diseases are caused by mutant alleles of genes that fail to properly encode proteins necessary for the regulation of cell functions, which in turn adversely affect the functions of different organs and organ systems in humans.

Some believe that Watson's goal soon will be realized. This is in the wake of the joint announcement in June 2000 by Craig Venter of Celera Genomics and Francis Collins of the National Human Genome Research Institute that they had completed a first draft of the entire mapped and sequenced human genome. Presumably, it will be only a matter of time before genetic testing can be used to predict whether people will likely develop disease over the course of their lives and thus serve as a form of disease prevention. Genetic knowledge could lead to better diagnosis of and treatment for diseases people have already. Collins asserts that the complete mapping and sequencing of the genome will make possible "a new understanding of genetic contributions to human disease and the development of rational strategies for minimizing or preventing disease phenotypes altogether."[6] This picture is misleading, however.

Diseases with a genetic cause may be divided into three types: monogenic, polygenic, and multifactorial. The first type is caused by a mutation in a single gene (cystic fibrosis, sickle-cell anemia, Huntington's disease). The second is caused by mutations in several or more genes (some cancers). And the third is caused by the interaction between several or more genes and environmental factors that trigger mutations in these genes (most cancers, heart disease, mental disorders). In fact, it would be more accurate to collapse polygenic and multifactorial diseases into one category, since genes and the environment together play a critical causal role in all diseases of the second and third types. Cancers, for example, usually result from multiple genetic mutations that are triggered by such environmental insults as radiation or various pathogens. Because these insults rarely occur all at once, cancers typically develop over an extended period of time and do not affect people until later in life. While many diseases with a genetic component are inherited as monogenic, there may be variations in the mode of inheritance that determine the risk of getting a disease, as well as the time of onset and its severity. In particular, a mutation in the BRCA1 or BRCA2 gene gives a woman who inherits it a lifetime breast cancer risk of between 50 and 85 %, as distinct from a 12% risk for the general population, where breast cancer appears to be multifactorial. Furthermore, a strong hereditary factor has been implicated in early-onset Parkinson's disease, while the later-onset and less severe version of this disease appears to be multifactorial.

Significantly, only a small proportion of the population is affected by monogenic, Mendelian, disorders. This is why claims that genetics will revolutionize medicine by focusing narrowly on specific genes are greatly exaggerated. To be sure, mapping and sequencing the entire human genome will lead to the identification of more genes implicated in various disorders. But most diseases are caused by more than the action of mutations in single genes. Neil Holtzman and Theresa Marteau attribute this to "the incomplete penetrance of genotypes for common (multifactorial) diseases."[7] In simplified terms, there is rarely a direct causal link between a given genetic mutation and a given disease. Having the mutation does not necessarily mean that one will develop the disease associated with it. Even in a family with a history of a heritable disorder, there may be varying degrees of penetrance, and so not every member of the family will develop the disorder. This underscores the ambiguity of the genetic component in disease. The proportion of people carrying the mutation who actually develop the disease is quite low. Holtzman and Marteau focus on the search for susceptibility-conferring genotypes for breast cancer, colon cancer, early-onset Type 2 diabetes, and Alzheimer's disease and point out that in each of these disorders genotypes account for less then 3% of all cases. They explain this by saying that "the risk of disease conferred by alleles at one locus depends not only on alleles at other, independently segregating loci, but also on environmental factors."[8]

Perhaps the best explanation for the incomplete penetrance of genotypes for disease is the phenomenon of *epigenesis*. Although the term itself literally means "above" or "over" genes, epigenesis is the science of heritable changes in gene expression that occur without a change in DNA sequence.[9] In this process, mutations occur not because of changes to the base pairs constituting the DNA molecule, but because of other chemical processes. Gene expression can be modified by chemical groups attaching themselves to the base pairs, affecting the functions of the proteins the genes encode and in turn the functions of the cells these proteins regulate. Methyl groups, composed of one carbon and three hydrogen atoms, produce such an effect through the process of methylation.

As Holtzman and Marteau further point out, "the complexity of the genetics of common diseases casts doubt on whether accurate prediction will ever be possible. Alleles at many different gene loci will increase the risk of certain diseases only when they are inherited with alleles at other loci, and only in the presence of specific environmental and behavioral factors."[10] The upshot is that the goals of better disease prevention and treatment will more likely be realized, not by a narrow focus on particular genes, but instead by studying genes within a broader causal framework that also includes behavioral, environmental, and other biological factors in addition to genetics.

Still, a satisfactory account of health and disease in medicine can be given only within an even broader evolutionary biological framework. This is if one accepts Randolph Nesse and George Williams's elaborated version of Dobzhansky's dictum: "Evolutionary biology is, of course, the scientific foundation of all biology, and biology is the foundation for all medicine."[11] The relations among evolutionary biology, medicine, and disease can be illustrated by a brief description of the function of the "innate" and "adaptive" components of the human immune system.

The innate component does not require prior sensitization by a specific foreign antigen to become activated. Rather, the elements of this system (macrophages, natural killer cells, complement, and naturally occurring antibody) are activated nonspecifically by any antigen. In the adaptive component, a period of prior sensitization to a specific antigen is required before it responds. Once sensitization occurs, memory cells are generated and remain in the circulation and lymphoid tissue for life. B cells and helper and cytotoxic T cells constitute the adaptive component of the human immune system. These two components and their distinct but related functions have been naturally selected in order to protect our organisms from potentially life-threatening microbes so that they do not become pathogenic. The memory cells in the adaptive arm of our immune systems enhance its ability to ward off infectious agents by enabling millions of molecular receptors to repeatedly recognize these agents and guide the body's defenses in the appropriate way.[12] Immune memory cells are a major reason why human organisms can survive despite constant threats from microbes.

Crucially, there must be a balanced response from the innate and adaptive arms to foreign agents. They must be potent enough to protect our organisms from virulent microbes, but not so potent that they cause autoimmune disease, which in extreme cases can lead to the demise of our organisms. This balance in the response of the two components of the human immune system has evolved to protect human organisms from infectious agents and thereby enhance their survival and reproductive fitness so that genes can be passed on to future organisms. Beyond the age of reproduction, however, natural selection has no reason to protect us in this way. We become more susceptible to chronic diseases as we age because our immune systems are not designed to work as effectively or efficiently later in life as they are earlier in life against acute infectious diseases. With the example of the immune system in mind, when questions of treating or preventing disease through gene therapy or other forms of genetic intervention arise, they should be placed within a framework of evolutionary biology, as this is the best way to understand how genes function in maintaining health or causing disease.

I have been using "health" and "disease" very loosely. Because they figure prominently as core terms in this discussion, some general definitions are in order. "Health" is the biological condition of the human body and brain when cells, tissues, organs, and organ systems function normally. In contrast, "disease" is the biological condition of the body and brain in which these same physiological features do not function normally, often because of the lack or dysfunction of enzymes or other proteins critical to the regulation of cellular processes.[13] Or to adopt Norman Daniels's view, "disease and disability are seen as departures from species-typical normal organization or functioning."[14] Daniels further says that "according to the normal functional model, the central purpose of health care is to maintain, restore, or compensate for the restricted opportunity and loss of function caused by disease and disability."[15] I will return to this definition of disease in chapter 3, when I discuss genetic enhancement and the goals of medical treatment.

Some diseases are both biological and mental at the same time. For example, schizophrenia is characterized as a disease of the brain involving dysfunction of neural connectivity and the neurotransmitter dopamine. It likely is caused by a sequence of multiple "hits," or insults to cell DNA, which include a combination of inherited genetic factors and nongenetic factors affecting the regulation and expression of genes controlling brain function.[16] In this respect, its etiology is similar to that of most cancers. Yet schizophrenia is clinically expressed as a disease of the mind involving cognitive and affective abnormalities. So, in this disease at least, the biological and the mental are intertwined.

Although I am in general agreement with Daniels's definition, disease must be distinguished from illness and disability. "Illness" is defined as the psychological experience of pain or other manifestations of disease. It is associated with suffering in the sense that illness involves one's subjective response to disease.[17] Thus, while the health or disease of the body or brain is defined in biological terms and refers to a human organism, illness is defined in psychological terms and refers to a person's unique phenomenological experience of living with a disease. A person may have a disease or illness yet not be disabled by it. "Disability" implies a general inability to formulate and pursue a life plan, where a mental or physical condition presents obstacles to undertaking and completing projects within that plan. In most cases, people are disabled by disease; but disease does not necessarily imply disability. If a woman has become infertile as a result of endometriosis but has no desire to have children through normal reproductive means and can control any pain associated with it, then this disease would not count as a disability for her. Moreover, hypertension may be classified as a disease in the sense that it involves a dysfunction of the mechanisms controlling blood pressure within the body. But if one is able to control both systolic and diastolic pressure within safe limits by taking appropriate antihypertensive medication, then one could live with the disease without becoming disabled by a stroke or cardiovascular disease.

Some might challenge the definition of disease in terms of species-typical functioning, claiming that there is no objective, value-free conception of function and therefore no objective, value-free conception of disease. Against the objectivist, whose conception of disease is grounded in facts about human physiology, the constructivist says that "function" and "disease" are social constructs formed relative to the values of different social groups or cultures.[18]

But constructivism mistakenly identifies the physiological functions that the body's organs and systems (cardiovascular, endocrine, immune, nervous, and so on) naturally are designed to do with how different social groups or cultures assign meaning to these functions in the way they affect the quality of people's lives. What must be distinguished, then, are the physiological dysfunctions that cause disease, on the one hand, and social judgments about quality of life given these dysfunctions, on the other. Nothing in these judgments vitiates the objective physiological definition of disease. More important, though, when we justify a medical intervention to treat or prevent a disease, it is by appealing to the way in which the disease affects or would affect people's quality of life. And there are some diseases entailing so much pain and suffering as to lead to a general consensus that the lives of people with these diseases would not be worth living. Prenatal testing and selective termination of early-stage fetuses or embryos with genetic anomalies causing these disorders could be justified on grounds of disease prevention. Intervention in the form of gene therapy might be justified for people who already exist with severe disorders, though the risks have to be weighed against the potential benefits.

In the light of these points, it is instructive to consider Sanfilippo syndrome. This is a genetic disease manifesting in early childhood and causing both severe mental retardation and aggressive behavior. As Philip Kitcher reasons:

> Testing to see if a fetus bears an allele for Sanfilippo syndrome is justified at bottom because the lives lived by these children with these alleles are sadly truncated and may diminish the quality of others, *not* because Sanfilippo is a disease or because people in Western society do not value children whose physical, cognitive, emotional, and behavioral development is massively disrupted. Deciding which types of prenatal testing or which types of molecular intervention are acceptable requires us to ask how the test and interventions would affect the quality of future lives.[19]

Returning to evolutionary biology, some authors estimate that up to 80% of all cancers are environmental in origin, suggesting that external insults to DNA are the main cause of this and other diseases.[20] This is a plausible explanation as far as it goes; but it does not go far enough. For an evolutionary explanation indicates that there is more to the multifactorial account of disease. On the multifactorial model, the more carcinogens or other external insults to DNA to which a woman is exposed, the higher will be her risk of getting breast cancer. This is supported by the fact that in the United States there is a correlation between the distribution of toxic-waste dump sites and regions of high breast cancer mortality. Presumably, the incidence of breast cancer would be much higher in these regions than in others. However, a study by the National Center for Health Statistics shows that the incidence of breast cancer is just as high in major metropolitan areas, especially in the Northeast, where toxic pollution levels generally are lower than elsewhere and where education and socioeconomic levels of women are higher than in the rest of the country. In contrast, the Southeast has a lower incidence of breast cancer while having a higher teenage pregnancy rate and a considerably lower proportion of college graduates and professionals than in the Northeast.[21] All of this suggests that an environmental explanation for the disease needs to be supplemented by an evolutionary one.

Higher education and income levels for many women in the Northeast indicate that increasing numbers of women are either foregoing or delaying having children until after they have established themselves in a career. If our bodies are designed by natural selection to ensure reproductive fitness early in life, then whatever disrupts this may adversely affect certain bodily processes, in this case the action of critical hormones such as estrogen on cells in breast tissue. We do know that childbearing before the age of thirty has a positive effect on the action of the relevant hormones. More women are having fewer babies, and are having them later in life. This means more menstrual cycles and consequently a greater amount of estrogen circulating in their bodies. Because of the effect of this hormone on cells in breast tissue in particular, not having children before age thirty puts women at an increased risk of developing breast cancer. An evolutionary explanation of the incidence of this disease is supported by the lower rate of breast cancer among women in the Southeast, where the rate of teenage pregnancy is higher. That is, women's bodies have been designed over many generations by natural selection to enhance reproductive fitness and success earlier in life. Behavior that does not accord with that design may make their bodies more susceptible to certain diseases, and delaying or foregoing reproduction can work against natural design and increase the risk of breast cancer. Over time, the body can adjust and become less susceptible to this disease when reproduction does not occur within a certain age range. But the period of time for the adjustment to take place would involve many generations of women.

Similarly, Nesse and Williams point out that, with respect to the human diet, the availability of certain foods has been so recent that natural selection has not yet had a chance to adjust to them.[22] This phenomenon has had adverse effects on the health of certain populations. For example, Native Americans traditionally had "feast-or-famine" diets depending largely on the availability of complex starches and lean meat from wild game. Now that foods consisting of saturated fats and refined carbohydrates are readily available to them, there is a much higher incidence of obesity and the associated diseases of hypertension and diabetes among them as compared with the rest of the general population. In evolutionary terms, the bodies of people in this particular population have yet to adjust to these dietary changes. Indeed, in many respects our bodies and physiology have not transcended our evolutionary past but rather are the products of that past.[23] Still, an evolutionary explanation cannot tell us what the causes of disease are in specific cases. It provides only a general understanding of the origins and development of disease in the human species.

Distinct from the three types of disease with a genetic component are three possibilities for a given genotype, the combination of alleles at a particular gene locus: normal, carrier, and disease.[24] In a normal genotype, there are at least two functioning copies or alleles of each gene, which means that the proteins the genes encode and the functions the proteins regulate will operate properly. With a carrier, there is one functional and one defective allele. By itself, the defective copy of the gene will have no adverse effects on cellular or other physiological functions. But if a carrier mates with another carrier and their alleles combine, then there is a significant probability that some of their children will inherit the mutation and develop disease. This scenario is typical of the mode of inheritance of sickle-cell anemia (SCA) and cystic fibrosis (CF), which are single-gene autosomal recessive disorders.

SCA is a disease caused by a mutation in the allele coding for hemoglobin, the oxygen-carrying protein in red blood cells. Two abnormal copies of the allele (one from each parent) cause the individual's red blood cells to become deformed, leading to blockage of blood vessels throughout the body. This results in acute pain and disability early in life and often an early death. CF is caused by an abnormally functioning protein, CFTR (cystic fibrosis transmembrane conductance regulator), which regulates the transport of chloride ions across cell membranes. This leads to excessive mucus production in the lungs, making one vulnerable to opportunistic infections caused by such bacteria as *Pseudomonas aeruginosa*, as well as the dysfunction of digestive enzymes in the pancreas, which greatly impairs the ability to digest food. Treatments for the symptoms of CF have improved over the years, but those afflicted with the disease usually die by the third decade of life and have decreasing quality of life as they age with the disease.

In recessive diseases, if a child inherits only one copy of a defective allele from one carrier parent, then he or she at most will be a carrier and will not actually develop the disorder. In fact, carrying one defective copy of a gene that would otherwise cause disease if it were inherited with a second defective copy can protect one from other diseases. Perhaps the most familiar disease fitting this description is SCA, where one copy of the sickle-cell allele not only does not cause the disease but also confers protection against malaria. This is prevalent among people native to equatorial Africa, where malaria has been an endemic disease. Interestingly, the relation between SCA and malaria seems to confirm the evolutionary hypothesis of the causal interaction between genes and the external environment regarding the incidence of disease. The allele protecting against the *Plasmodium* responsible for malaria is selected for in an environment where this protozoan predominates, thereby enhancing the survival of human organisms in that environment.

Most inherited conditions are autosomal recessive. This means that if two carriers have children together, then each child has a one-in-four chance of inheriting the condition and having the symptoms associated with it. A smaller number of conditions are of autosomal dominant inheritance, which means that each child of a parent who carries a trait for a given condition will have a one-in-two chance of inheriting it. These disorders are rare because the people who have them are often too sick to have children, though people with the late-onset Huntington's disease may already have children before knowing that they have the disorder. With dominant conditions, only one copy of a defective allele needs to be inherited for an individual to have the genetic disease it causes. So, while recessive genetic conditions are more frequent than dominant ones in the human population, the probability of a parent passing on a dominant trait to offspring is greater. Moreover, although recessive disorders generally are more severe than dominant ones, the latter can be severe as well, as in Huntington's. Equally serious are disorders traceable, not to any of the twenty-two autosomes, but instead to the sex chromosomes, X and Y. Well-known sex-linked disorders are Duchenne muscular dystrophy and hemophilia, both of which are traceable to mutations on the X chromosome and are inherited from the mother by half of the male children she bears. There are also disorders traceable to chromosomal anomalies that are not inherited but develop during fetal gestation, specifically trisomies 13, 18, and 21 (Down syndrome).

In chapter 2, I will discuss the prudential and moral importance of the differences in probabilities of inheriting dominant and recessive traits. I also will consider what the probability of inheritance means for the severity of a disorder, the age of onset of symptoms, the quality of life of affected individuals, and the projected life span. In particular, I will discuss what our moral obligation is to these individuals if we decide to bring them into existence and whether we have an obligation to prevent them from existing at all. In addition, there will be more detailed descriptions of the biological or medical characteristics of some of these monogenic diseases when I examine cases of people whose lives are or would be defined in terms of them. I also will address obligation and other moral issues with respect to polygenic or multifactorial diseases, where the genetic component is not the only medically or morally relevant consideration.

Biological and Personal Identity

The particular genotypic and phenotypic traits that physically distinguish each of us from other human organisms constitute our biological identity. But the biological identity of a human organism is not equivalent to the psychological identity of a person. The mental states in virtue of which we are consciously aware of ourselves as persisting through time depend on the continued normal functioning of certain regions of the brain. These include the cerebral cortex, which controls reasoning and other thought processes, the hippocampus, a center of memory, and the thalamus, which organizes sensory images to and from the cerebral cortex, among others. Our mental life also depends on the various neurotransmitters and hormones in the nervous and endocrine systems that regulate our cognitive, volitional, and affective capacities. The body plays an important role in our mentality as well, serving as a ground reference for the conscious representation of ourselves as entities located in the world and as a link to the somatosensory system in the brain.[25] Still, mental states are not reducible to, in the sense of being completely explained by, the brain and body. Consciousness is a property that emerges from the normal functioning of the relevant regions of the brain, but qua emergent is not identical to these regions or their properties. Furthermore, the qualitative character of consciousness—*what it is like* to perceive colors or remember an experienced event—makes it a psychological property that cannot be explained entirely in biological terms.[26] I have noted the role of the body in our conscious representations of ourselves as persisting through time. But the contents of these representations, and indeed of our mental states in general, are at least partly determined by features of the social and physical environments in which we exist. Personhood and personal identity are psychological concepts. But the nature and content of the mental states in which personhood and personal identity consist causally depend on the brain, body, and the social and physical environment.

According to the *Oxford English Dictionary*, a "person" is "a self-conscious or rational being." This is similar to the definition given by Derek Parfit, who claims that "to be a person, a being must be self-conscious, aware of its identity and continued existence over time."[27] The criteria of being a person, which consist of mental states caused by certain brain and bodily functions, are more complex than the criteria of being human, which involve brain and bodily functions but not those necessary to generate and sustain mental life. This implies that persons are not identical to human organisms, or human beings. Although the two terms sometimes are used interchangeably, "person" is a psychological concept, while "human being" is a biological concept. As I will explain shortly, there are good reasons for insisting that we are essentially persons rather than human organisms.

Being a person at a time is closely related to persisting as one and the same person through time. Personhood is a synchronic concept, while personal identity is a diachronic concept. Following Parfit and others, personal identity consists in the holding of relations of psychological connectedness and continuity between and among mental states over periods of time.[28] Because the relations among mental states and events necessarily depend on the brain and body, and because no entity such as a soul or Cartesian ego is presupposed as existing prior to the holding of these relations, the position I am adopting rejects dualism. This theory says that while there are correlations between mental properties and physical properties in the body or brain, these correlations are contingent rather than necessary. For dualists, the mind does not depend essentially on the brain. The position I am adopting also rejects materialism, which says that the mind, and therefore personhood, is nothing more than physical processes occurring in the brain. For materialists, mental states are not simply *caused* by neural processes; they *are* neural processes. Yet because mental states have contents deriving from the social and physical environment, and because they can have a qualitative character, they cannot be explained entirely in terms of biochemical or physical processes in the body and brain. So the psychological connectedness account of personhood and personal identity that I endorse rejects both materialism and dualism.[29] The position I take is similar to what Parfit calls the "Reductionist View." In his words: "On the Reductionist View, each person's existence just involves the existence of a brain and body, the doing of certain deeds, the thinking of certain thoughts, the occurrence of certain experiences, and so on."[30] Provided that we locate them within a social and physical environment, persons do not exist apart from the physical and mental events and states in terms of which we identify them. If this is a form of reductionism, then we should call it broad psychological reductionism.

According to Parfit, psychological connectedness consists in the holding of particular direct links between mental events, such as the persistence of beliefs and desires, the connection between an intention and the later act in which it is carried out, and between an experience and one's memory of it. These connections can be stronger, holding over shorter periods of time, or weaker, holding over longer periods. Psychological continuity is the ancestral relation of psychological connectedness, consisting of overlapping chains of strong connectedness and extending over longer periods of time than what is involved in particular links between mental states.[31] Unlike connectedness, continuity does not admit of degrees and is a transitive relation. That is, it embodies the truth of the syllogism: If A at T1 is identical to B at T2, and if B at T2 is identical to C at T3, then A at T1 is identical to C at T3.

The idea that the connections between mental states hold only over limited periods of time has significant implications for the idea of prudential concern about one's future self in a genetically manipulated extended life span. I will explore these implications in chapter 5. For present purposes, though, I need to defend the distinction I have drawn between persons and human organisms, as well as my claim that we are essentially persons.

A person is a psychological kind in the sense that it is defined essentially in terms of the psychological properties associated with mental life. A human being is a biological kind, an entity consisting of an integrated set of biological structures and functions. At a very early stage of development, these structures and functions include the potential for and eventual actualization of a brain stem and autonomic nervous system to sustain heart rate, respiration, and the functions of other organs of a human body.[32] A functioning cerebral cortex and other supporting structures are not necessary for something to be a human organism. In its early stage, a human organism has the potential to develop these regions of the brain necessary for mental life. Yet even if this potential is never realized, the organism can continue to exist. Because it possesses only the potential and not also the capacity for mental life, and because the structures and functions of its cells, tissues, and organ systems can be explained entirely in biological terms, a human being is a biological and not a psychological kind. "Capacity" implies the ability of an entity to exercise some function at any given time, which is not implied by "potential." And although some may use "human organism" to refer to a zygote, embryo, or early-stage fetus and "human being" to refer to a full-fledged individual with a body and (part of) a brain, these can be considered different stages in the development of one biological entity. Thus I use "human being" and "human organism" interchangeably.

Given that the capacity for consciousness and other forms of mental life that define personhood are generated and sustained by certain biological features of human organisms, persons are closely related to their organisms. But these two types are ontologically distinct in that the capacity for mental life is something that only persons possess. A being can be human without ever having this capacity, in which case it never will be a person, or else having permanently lost this capacity, in which case it no longer is a person. Put another way, a person is constituted by a human organism but is not identical to that organism.[33]

An entity that has a functioning brain stem and certain functioning internal organs, yet lacks a functioning cerebral cortex and other structures supporting mentality, would be a human being but not a person. For example, an anencephalic infant would on most accounts be considered a human being but not a person because it lacks the cerebral cortex necessary for consciousness and other mental states. Similarly, an individual in a persistent vegetative state who has permanently lost the capacity for mentality also would be considered a human being but not a person. If we take ourselves to be essentially entities with the capacity for consciousness and other forms of mental life, if this capacity is definitive of personhood, and if being human does not entail this capacity, then it follows that we are essentially persons rather than human organisms or human beings.

To further support this conclusion, I will spell out two distinct senses of "life" corresponding to persons and human organisms, respectively. The fact that these two senses are not coextensive and that one begins before and ends after the other will reinforce my claim that persons are ontologically distinct from human organisms. It should be noted, though, that I am elaborating this distinction with a view to examining how genes and different forms of genetic intervention can shape or alter the identity of persons. Genetic identity is not personal identity. But the manipulation of genes at different stages in the life of a human organism or a person can affect the nature of and connections between mental states and over time determine whether a given set of mental states is that of one person or distinct persons.

Two Senses of "Life"

It will be helpful to introduce some metaphysical terms to elucidate the issues at hand. Metaphysicians often employ sortal concepts to establish criteria of identity for entities. These criteria determine what counts as being a thing of a certain sort, as well as what the continued existence of things of that sort necessarily involves. They distinguish between *phase* sortals and *substance* sortals.[34] A phase sortal designates the sort or kind to which an individual belongs through only a part of its history. "Child" is a phase sortal, since although one was not a child when one began to exist, one later became a child and then ceased to be one, all the while remaining one and the same individual throughout the various transformations. A substance sortal, in contrast, designates the sort or kind to which an individual belongs throughout its entire existence. Substance sortals indicate the sort of thing an entity essentially is, the sort of thing it cannot cease to be without ceasing to exist. Regarding the question of what we essentially are, it is clear that the concern is with substance sortals. The fundamental question is whether the relevant substance sortal is the psychological "person" or the biological "human organism," and I have been arguing that the answer is the former.

My main claim and argument on this question can be supported by analyzing when our psychological and biological lives begin and end. For simplicity, I will stipulate that the life of an entity is the period of time through which we can trace and identify it. If we are essentially persons defined in terms of the capacity for mental life, then our lives begin much later than conception. According to the psychological connectedness/continuity account of personal identity, a person does not exist until the connectedness of mental states from day to day is strong enough to generate self-consciousness. This suggests that even in early infancy there is no self-conscious entity present and therefore no person exists because the requisite psychological connections do not yet hold. Some even argue that one cannot become self-conscious until one is aware of the social and physical environment around one, in which case one would not become a person until at least several years after birth.[35] Significantly, self-consciousness does not appear all at once but gradually, thus implying that a person comes into existence gradually. Because this is a gradual process, the question of precisely when a person comes into existence does not have a determinate answer. Nevertheless, we can say that the rough time at issue must be after the brain has developed the necessary structures and functions to generate and support the capacity for mentality. At the earliest, a person begins to exist in the final stage of fetal development, but more plausibly in infancy. At the other end of psychological life, a person ceases to exist when the cerebral cortex and other supporting regions of the brain permanently cease to function. Yet because the functions of these regions usually diminish gradually, in most cases persons cease to exist not all at once but gradually. The question of precisely when a person ceases to exist lacks a determinate answer as well.

If we are essentially organisms, then some would claim that we begin to exist at syngamy, when the genetic materials from the genetic father's sperm and the genetic mother's egg fuse to form a single-cell zygote. We cease to exist when all functions of all parts of the brain, heart rate, and respiration permanently cease. In neither case is there a precise time at which a human organism begins or ceases to exist. Syngamy is not an event but a process that is not complete until about twenty-four hours after fertilization.[36] Others might claim that as human organisms we begin to exist around fourteen days after syngamy, once the possibility of monozygotic twinning has passed. This involves the zygote dividing from within to form two qualitative identical zygotes. These are continuous with the original zygote but numerically distinct from the original as well as from each other. So there is no single entity that persists throughout the process that begins with fertilization. If we are essentially human organisms that begin to exist at syngamy, then when the zygote phase of our organisms divides into two organisms, one of us ceases to exist. But since there are now two zygotes and thus two organisms that exist following monozygotic twinning, human organisms cannot begin to exist at syngamy but at the earliest some fourteen days or so later.

If I were essentially an organism, then my life would have begun when I was a embryo. But this implies that I once existed as an undifferentiated clump of some eight or twelve cells, and it is unclear how anyone could trace and identify *me* among these cells. At the other end of the spectrum, if I were essentially an organism, then I could irreversibly lose my capacity for mental life and still continue to exist. I would continue to exist with only minimal activity of the brain stem regulating heart rate and respiration. Yet it seems implausible that, if I were in a persistent vegetative state with an irreversible loss of higher-brain function, then I would continue to exist. The biological life of our organisms may continue after our psychological life has ended. But it is our psychological lives, our lives as persons, which we take to be essential to us. For all of the reasons I have adduced, we should reject the view that we are essentially human organisms, or human beings, and maintain that we are essentially persons. Again, persons are constituted by organisms but are not identical to them. The life of a person begins later than and ends earlier than the life of a human organism, even though we cannot specify an exact time when it begins or ends.

Let us suppose that a human organism begins to exist in the form of a zygote around fourteen days after conception and that this will develop into an embryo, fetus, and eventually a person. While an increasing number of cells follow a certain line of development in multicellular organisms like humans, this does not mean that a single individual exists throughout the entire process. For during this period of development, embryonic cells are undergoing a continuous process of differentiation into distinct cell, tissue, and organ types. Differentiation implies that biological properties undergo significant qualitative changes from the stage of the zygote and embryo to later fetal stages. As embryologist C. R. Austin notes, "the whole embryo *does not* become the fetus—only a small fraction of the embryo is thus involved, the rest of it continuing as the placenta and other auxiliary structures."[37] Moreover, even if the cerebral cortex is present in rudimentary form in fetuses and has the potential to generate mentality, this does not mean that fetuses become (in the identity-preserving sense of "become") persons. As molecular biologist Lee Silver points out, "although the cerebral cortex—the eventual seat of human awareness and emotions—has begun to grow, the cells within it are not capable of functioning as nerve cells. They are simply precursors to nerve cells without the ability to send or receive any neurological signals. Further steps of differentiation must occur."[38] All of this suggests that one's genotype is not identical with one's phenotype, and that the capacity for mentality which makes us persons is not completely determined by either of these biological types.

Earlier, I noted that incomplete penetrance and epigenesis account for the fact that most diseases cannot be explained in terms of a direct causal connection to a particular gene. A second feature of epigenesis is that it also accounts for the fact that the identities of persons cannot be attributed to the functions of genes alone. Successive stages of differentiation in the course of an organism's development give rise to new structures with new properties. The genetic code in the zygote or embryo only controls the *general* range of possible phenotypic outcomes in a human organism. It does not determine which *particular* traits will emerge at the end of the process. This is because the genetic code itself cannot account for the interactions between the products of successive stages of development, and because the translation of genetic potential into a particular phenotype depends on the environment in which development takes place.[39]

Since we are considering development from zygote to embryo to fetus to full-fledged human organism, it is the uterine environment that is especially pertinent here. With regard to the brain, certain genes code for a range of synapses and neurotransmitter levels. But the particular synapses that form and the levels of the neurotransmitters mediating between the synapses result from the interaction between genes and the uterine environment during fetal development. That factors in this environment can adversely affect the potential of genes coding for certain brain structures and functions that otherwise would be realized in a normal way is evident in anencephaly and schizophrenia, though of course the nature of the multifactorial causes is different in these two conditions. If genes by themselves do not determine the particular structures and functions of our brains and bodies, and if our brains and bodies do not by themselves determine the content and phenomenological character of the mental states that make us persons, then personhood and personal identity are influenced but not determined by genotype and phenotype.

The upshot of this account of cell differentiation is that the mere potential of one stage of a human organism to develop into another stage is not enough to establish a relation of identity between these stages. Thus a person never was an embryo or fetus, and embryos and fetuses do not become persons. An embryo or presentient fetus is a potential person, not in the sense that it becomes a person, but only in the sense that it has the potential to develop the biological structures and functions necessary to generate the capacity for consciousness and other mental states that define persons.[40] When the life of a person roughly begins will be critical to assessing the effects of genetic manipulation on personal identity, as well as to whether and to what extent the recipients of this manipulation can benefit or be harmed by it.

How Genes Influence the Identities of Persons

I have argued that the genetic properties of a single-cell zygote and early embryo are not identical to the biological properties of later embryonic and fetal stages of development of a human organism. For the genotype only determines the range of possibilities in which phenotypic traits are expressed, and the translation of genetic potential into these traits depends on the internal biological environment in which genes interact and cell differentiation occurs. The epigenetic nature of differentiation means that new and distinct biological properties will emerge from one stage of development to the next, a process that is not completely determined in advance by the genotype. I also have argued that neither genotypic nor phenotypic identity is equivalent to personal identity. Neither of these biological types that each of us possesses determines who we are, for two main reasons that are worth repeating. First, the phenomenological character of desires, beliefs, intentions, memories, and other mental states causally depends on but cannot be explained completely by the ways in which genes influence synapses and neurotransmitters in the brain. Second, the contents of our mental states are largely a function of the social and physical environment in which we exist, and these cannot be explained completely by genes or their products either. Nevertheless, it is important to articulate the respects in which genes and different forms of genetic intervention can influence personal identity.

The genotypic polymorphisms that result in our distinctive phenotypic traits such as hair and eye color, stature, physiognomy, and other physical features of our bodies play a critical role in our self-consciousness awareness. Again, the body serves as a ground reference or locus in which our mental states generated and sustained by the brain are unified over time. The body mediates between our desires, beliefs, and intentions and their contents in the external world. So there cannot be any disembodied mental states. Given that the mental states in which personhood consists are grounded in the interaction between the brain and the body, the condition of the body—whether it is diseased or healthy—can affect the contents of and connections between these mental states. Insofar as the condition of the body is at least partly the result of gene expression in physical traits, and the condition of the body influences the mentality definitive of personal identity, genes can at least indirectly affect personal identity. To make the point slightly differently, the biological condition of disease can affect one's sense of self in the way one experiences the disease as the psychological condition of illness.

For example, a person with an early-onset monogenic disease such as CF or SCA, which involve considerable pain and disability and a relatively short life span, likely would have a different set of desires, beliefs, intentions, and memories than a person who did not have the same disease. The experience of pain and disability would affect the qualitative aspects of the memory of one's experiences of the past, as well as how far into the future one's desires and intentions could extend. This in turn would affect the connections between the person's mental states, which would be qualitatively different and would be connected for a shorter period than those of a person whose entire biological and psychological life was free of disease and illness, until just before death.

The influence of genetics on identity is most pronounced in certain cognitive and affective disorders. In these cases, the very ability to have mental states and the conditions of psychological connectedness and continuity are directly shaped by the genes coding for critical proteins regulating brain biochemistry. Fragile X syndrome and Trisomy 21 (Down syndrome) are chromosomal disorders adversely affecting brain and mental function from birth, causing varying degrees of mental retardation and the existence of different persons from those who would have existed without these anomalies. Perhaps the most vexing multifactorial brain and mental disorder is one that I already have mentioned, schizophrenia. What sets this disorder apart from the two I just cited is that the delusions, hallucinations, and paranoia characteristic of it usually do not manifest themselves until late adolescence or early adulthood. This can result in radical disconnectedness and discontinuity between the mental states and capacities an individual has before and after the onset of symptoms. Indeed, the disunity between mental states can be so radical that there are effectively two different persons before and after the symptoms of the disease appear. These can wax and wane over varying periods of time. While environmental factors (e.g., the presence of a virus in the uterine environment during fetal gestation) seem to play a causal role in the etiology of schizophrenia, another cause of this disorder are mutations in several genes affecting dopamine. This and other neurotransmitters are gene products that regulate the synaptic connections in the brain controlling our cognitive, volitional, and emotional capacities. Hence the genetic component in this disease and its impact on the identities of the people who have it is significant.

It is important to distinguish between early- and late-onset genetic disorders with a view to how they bear on personal identity. This will take on special importance when considering the impact of gene therapy and other forms of genetic intervention on embryos, fetuses, and persons. Let us assume that a person does not exist until infancy or even early childhood, when one is conscious of oneself as an individual with a particular body, is aware of the social and physical environment, and has a minimally unified set of mental states. Once these connections have been formed and are strengthened over time as one lives longer, significant disruption in the connectedness and continuity of mental states would have to occur for identity to be altered. Intuitively, in a monogenic physical disorder like CF, even if gene therapy at some point in the individual's late childhood or adolescence restored normal lung and pancreatic function, it is doubtful that this cure would effect a change in the identity of the person. To be sure, one would be different in the sense that the contents of and connections between one's desires and intentions would change somewhat given the knowledge that one would live longer and would not be physically restricted by the disease. Yet one would not literally become a different person, assuming that the memories of one's past experience with the disease would not be lost but would remain intact. The connections between one's experiences and one's memories of them are one of the criteria of personal identity. Indeed, the backward-looking memories of pain and suffering with the disease may very well shape the forward-looking desires and intentions without the disease. The significance of these remarks hinges on the idea that gene therapy holds the only promise of a cure for monogenic diseases.

In the case of an older child, adolescent, or adult receiving gene therapy for a disease like CF, the individual's biological life after the procedure would be distinct from their biological life before the procedure. But unless the body were radically transformed as a result, there would be one person with one continuous psychological life before and after the therapy. In Parfit's words, "to say that literally a different person appeared would require a break in the deeper relation of continuity, not just connected-ness."[41]

The retention of memories of events experienced when one had the disease would imply that continuity would remain intact. As a set of mental states becomes more integrated and unified over time, the more radical the changes to the connectedness and continuity between these states must be to disrupt them and alter personal identity.

The earlier genetic intervention to prevent, treat, or cure a physical disorder occurred in one's biological life, the more likely it would be identity-determining rather than identity-preserving. That is, it would determine that a distinct individual would exist following the intervention than the one who existed before it, or the one who would have existed without the intervention. If a gene or genes could be deleted, added, or corrected at the embryonic or early fetal stage of development, then the intervention would occur while cell differentiation was still underway. To the extent that the alteration of the gene would affect other genes and gene products through epigenesis, it would alter the range of possible outcomes specified in the zygote. A different sequence of cell differentiation would take place at successive stages of biological development. Consequently, there would be a different set of physical traits than those that would have emerged if no intervention had taken place. Insofar as the psychological properties definitive of personhood depend on these physical traits, genetic intervention that occurs while cell differentiation and epigenesis are still in progress determines these traits and thus determines that a different person comes into existence from the one who would have existed otherwise. At the point of intervention, no psychological connections could have been formed yet and therefore no person could exist at this early stage of development. Intervening at the embryonic or early fetal stage of development would determine distinct biological and psychological lives from the lives that would have existed without the intervention. More specifically, if we trace the biological life of the individual with CF back to its beginning, when the gametes from the two carrier parents of the CF allele produced the zygote from which the child developed, correcting the mutation on one of these alleles would mean that they would have a child who was a carrier of the trait, not a child who would develop the disease. The identity of the child would be distinct in each of these two scenarios.

In contrast, if gene therapy could cure an individual with CF or SCA during late childhood or adolescence, then from the time of the therapy we would be considering a distinct biological life, but the same psychological life of the same person who existed before the therapy. Unless the alteration involved genes coding for proteins regulating synapses and neurotransmitters in the brain, thus directly affecting the nature of one's mental states, or else entailed a radical transformation of one's body, the continuity between mental states before and after the procedure would not be disrupted. As I have noted, this is due to the fact that memories of past experiences are partly definitive of personal identify, as are intentions regarding the future, which are shaped by these memories. A person likely would retain a stable set of memories if she were cured of a physical disease through somatic-cell gene therapy.

A case that illustrates this point is that of Ashanti de Silva, the first person to be treated successfully for a genetic disorder with gene therapy.[42] She was afflicted with ADA-SCID, severe combined immunodeficiency, due to a mutation in the gene coding for the enzyme adenosine deaminase (ADA), which regulates the metabolism of one of the nucleotide bases of DNA. ADA-SCID has a recessive mode of inheritance, requiring two defective alleles, one from each parent. In this disorder, mutations in several genes thwart the production of T lymphocytes, making one highly susceptible to infectious disease. In 1992, at the age of five, a few of her T cells were withdrawn from her blood and injected in culture with a normal copy of the ADA gene, after which her modified T cells were introduced back into her bloodstream through an intravenous drip. This resulted in an amelioration of the symptoms of her disease. In 2001, at age fourteen, her T cells and immune system appear to be functioning normally.

Although the contents of some of Ashanti's future-oriented mental states (desires, intentions) are probably different than they would be if she experienced the severe symptoms of the disease, there is no indication that her past-oriented mental states (memories) have radically changed and that she has become a completely different person. Her biological and psychological lives likely will be longer and will be comprised of different events and experiences than those of the shorter biological and psychological lives with the full-blown disease. But all accounts suggest that the continuity of her mental states, while richer than before, has remained fundamentally the same over the last nine years and that she has remained, and likely will remain, the same person.

Consider now the successful treatment in early 2000 in France of two male babies, aged eleven and eight months, respectively, for an X-linked form of SCID.[43] The biological consequences of this intervention are significant because they mean that the babies will develop stronger immune systems and be less vulnerable to life-threatening infections. They will have longer biological lives as a consequence. Moreover, even though the intervention occurred after most cell differentiation had taken place, neither baby had yet developed an integrated set of psychological properties and thus there was no personal identity to be disrupted. The different biological properties of the disease- and symptom-free bodies the babies develop will entail a different set of psychological properties and therefore different persons from those who would have existed without the treatment. As distinct from the case of Ashanti de Silva, this case shows that the earlier genetic intervention occurs in the life of a human organism, the more likely it is to determine the identities of distinct persons than to preserve the identity of one person.

Moral Implications

Let us now briefly consider some of the moral implications of the relation between genes and personal identity. These will be examined in more detail in later chapters. But it will be helpful to introduce them at this point.

Only persons can have interests, since only persons have the mental capacity for desires, beliefs, emotions, and intentions that are presupposed by the very idea of interests. Insofar as benefits and harms result from the satisfaction and defeat of interests, only persons can be benefited or harmed.[44] A person is benefited when she is made better off than she was or otherwise would have been, and she is harmed when she is made worse off than she was or otherwise would have been. This requires a comparison between two states in which one and the same individual exists.[45] Since persons are not identical to the embryonic or fetal stages of the human organism from which they develop, and since embryos and fetuses do not have the requisite biological structures to generate and sustain the mental capacity necessary for interests, embryos and early-stage fetuses cannot be benefited or harmed. It follows that at no stage of development prior to the emergence of personhood and personal identity can a human organism benefit from, or be harmed by, any form of genetic modification.

Nevertheless, given that genetic modification at these stages subsequently can result in effects on people's bodies and minds, people at least indirectly can benefit from or be harmed by such modification. The people in question would not have any interests at the time the modification took place, since they would not yet exist. But insofar as the insertion, deletion, or correction of a gene, together with its interactions with other genes, can affect protein synthesis in cells and determine whether this results in disease or its prevention, people can be harmed or benefited by such genetic intervention. The benefit or harm occurs, not because before they existed the people in question had an interest in not having disease. Rather, once people exist, they have an interest in not experiencing the pain, suffering, and restricted opportunities entailed by severe disease and disability.

Furthermore, to the extent that whether a person is harmed in this way is controlled by those who bring him or her into existence, and to the extent that they know that there is a significant risk of a genetic anomaly or anomalies causing diseases in their offspring, parents may be responsible for whether a child is harmed from a disease during its life. In general, knowledge of the probable effects of a given action or omission gives one a certain measure of control over those effects, and control over effects is a necessary condition of being responsible for them. This assumes that one can take action to prevent these effects from occurring. Parents of a child with a genetic disease would be responsible for that condition only to the extent that they had the requisite knowledge and could have availed themselves of procedures to prevent having a severely diseased child. These could include genetic interventions of the sort I have described, or alternatively genetic testing and selective termination of the embryos with the genetic anomalies. In chapter 2, I will discuss whether the differences in physical and mental symptoms between early- and late-onset genetic disorders entail a difference in our evaluation of obligation, responsibility, and harm in bringing into existence people who will have these disorders.[46]

It seems that what matters here is not so much *who* will or would have to experience the symptoms of genetically caused diseases, but rather *that* some future people will or would have to experience them. That is, if we were to exploit genetic technology and intervene at some point in the life of a human organism or person to correct a genetic mutation and thereby prevent, cure, or ameliorate the symptoms of a disease caused by that mutation, then presumably *when* we intervene and *whom* it affected would not be significant as *that* we intervene. The point is to prevent the disease and its symptoms so that *no one* will have to experience its harmful effects. The moral question of whether the genetic code in a zygote or embryo, or manipulation of that code along the pathway of biological development, can benefit or harm the people who result from it seems more important than the metaphysical question of who these people are or will be. But if benefit or harm consists in making a person better or worse off, then we must identify the person who stands to be made better or worse off. Hence the moral question of benefit or harm cannot be separated from the metaphysical question of who is benefited or harmed. The two questions are inextricably intertwined.

If we use gene therapy to correct a mutation or add a normal copy of a gene to a person's somatic cells, then we must exercise caution in the delivery of the gene or genes in question. This is because we still lack precise knowledge of how different genes causally interact with each other. Altering one mutant gene may adversely affect other genes and in turn the functions of critical enzymes and other proteins in cells. Unfortunately, the general method of delivering genes into cells through viral vectors does not always affect only the targeted genes with mutations but others as well, with possible untoward effects on different cellular and bodily functions. Furthermore, if we genetically alter germ cells in the sperm and egg, then the effects will not remain with the affected person but will be passed on to future people. Because we cannot be sure how altering one gene will affect the other genes and the proteins for which they code in regulating normal bodily functions, and because unintended and unforeseeable cellular and bodily dysfunctions could result that would be passed on to future people, we have an obligation to proceed with caution on this front.

Genetic information can be derived from testing cells of individuals (preimplantation IVF embryos, fetuses, or persons) who, because of the presence of inherited traits in their families, are at risk of carrying a recessive or dominant allele that causes a serious disease. It also can be derived through genetic screening of children, adolescents, or adults in wider groups or populations known to have a significant incidence of mutations causing disease. These tests are extremely powerful tools that can predict a person's future health status and lead to the prevention of diseases to which people may be predisposed. This would not come through genetic intervention but through diet and behavioral modification, given one's genetic profile.

The Institute of Medicine Committee on Assessing Genetic Risks has recommended that genetic testing only be allowed for disorders for which a cure or preventive treatment exists. Accordingly, they recommend that screening is not appropriate for detecting carrier status, untreatable childhood diseases, and unpreventable late-onset diseases.[47] But surely testing should be made available to yield information about untreatable and incurable disorders that a carrier or affected parent could pass on to a child. Knowing that there was a risk of this occurring would give a prospective parent some degree of responsibility for the health of their future child. This knowledge would enable them to prevent such a disorder by informing their reasons and choice for or against having the child. Insofar as we have an obligation to prevent disease in future people given their interest in not being harmed by disease, and insofar as the knowledge yielded through genetic testing can lead to the prevention of disease, parents may have an obligation to obtain this knowledge through testing. In addition to testing embryos for genetic anomalies and deciding whether to allow them to develop or to selectively terminate them, an adult intending to become a parent could be tested for carrier status and, if the test were positive, avoid having children with another known carrier. This would be one way of disease prevention without treatment. There are also prudential reasons for testing for untreatable late-onset diseases as early as possible. The information could allow a person who would be affected to plan the remainder of his or her life in a more meaningful way.

Another important moral issue is whether all people should have equal access to genetic technologies to prevent, treat, or cure diseases or, more controversially, to enhance normal physical and mental traits. For many people, the delivery of necessary or desired genes into somatic or germ cells through viral vectors or pharmacogenomics (drug therapy tailored to specific genes) would be prohibitively expensive. The morally objectionable consequence of this is that only people who could afford this technology would be able to avail themselves of it. With respect to disease prevention and health promotion, it would make some people better off than others who are not well off to begin with through no fault of their own. With respect to genetic enhancement, it would make some people who are already well off even better off, both in terms of an absolute baseline of normal health status and compared with those who are worse off than they in the same regard. This would exacerbate the problem of social and economic inequality and would raise deep questions about the sort of society in which we live and would want to live. I will examine this issue more closely in chapter 3.

In chapters 2 and 3, I will address a related moral question that arises when we compare the cost of genetic intervention, and the number of people with monogenic diseases who would benefit from it, with less costly preventive or therapeutic measures that would benefit a much greater number of people with multifactorial diseases. Which of these two groups has the stronger claim to medical care? This is especially significant given the fact that such single-gene disorders as CF, SCA, Tay-Sachs, and phenylketonuria (PKU) affect only a relatively small number of people.[48] On what moral grounds can we give research and treatment priority to the more severe diseases of fewer people over the less severe diseases of many more people? If our general aim in medicine and biotechnology is to prevent disease and promote health in the greatest number of people, then it is not so clear that we should spend a disproportionate amount of limited medical resources on genetic technology that might achieve this goal in only a small percentage of the human population.

Conclusion

I have explained how genes influence biological structures and functions in the development of human organisms. I have outlined the different types of causal role that genes play in the etiology of disease of these organisms and how this bears on the metaphysical question of personal identity and the moral question of whether persons can benefit from or be harmed by the genes in their body's cells or by alterations to these genes. At a deeper level, the ways in which genes influence cell and organ functions is attributable to the evolutionary force of natural selection, whereby alleles of genes are selected to enhance the survival of human organisms. The purpose of this is to confer a reproductive advantage on humans to enable them to transmit their genes into the next and successive generations. The behavior of genes can be understood only within a framework of evolutionary biology.

We cannot escape the impersonal force of natural selection, as is evident in the effects of genes on the physical and mental health of persons within particular generations. This is the result of genes evolving over millions of years. Still, this should not be taken as a pessimistic assessment of our biological fate. The force of natural selection operates at the glacial pace of evolutionary time, which is quite different from the time frame in which our present actions and their consequences occur. This allows us some elbow room to exercise some control over the respects in which genes affect our biological and psychological lives, as we now do on a relatively small scale with different forms of genetic intervention into the cells of our bodies. With this technology, we can to some extent determine whether people will exist with healthy or diseased lives, as well as the sorts of physical, cognitive, and emotional properties they will have. By the same token, this ability to intervene in genetic pathways entails a considerable degree of moral obligation to use it in such a way as to benefit and not harm people.

Indeed, genetic interventions to prevent or modify physical and mental capacities in existing and future individuals bear not only on the moral issues of benefit, harm, obligation, and responsibility. They also bear on the broader social and political issues of fairness and equality regarding access to this technology.

Whether our purpose is to prevent or cure diseases with a genetic component or to enhance normal traits, if we decide to manipulate genes in one's somatic cells, then the effects of the manipulation would remain with the individual undergoing the procedure. But if we decide to prevent an allele from expressing in a disease in future people by eradicating it at the germ line, then the effects if this procedure—beneficial or harmful—would be passed on to future generations without their consent. Because we lack complete knowledge of the complex interactions between and among genes and gene products, it is extremely difficult, if not impossible, to foresee the long-term effects of genetic modification. The sorts of genetic interventions we undertake now may adversely affect genetic diversity, which consequently might make the bodies of distant future people less able to adapt to changing environmental conditions. This could make them more vulnerable to pathogens and disease and less able to live out what we would consider a normal life span.

In the chapters that follow, I will not be recommending an absolute prohibition on all forms of genetic intervention, as some of them may yield benefits for existing people and those who will exist in the near future. However, our lack of adequate knowledge of how genes interact amongst themselves, how they influence biochemical processes in the body, and of whether manipulating them will adversely impact on genetic diversity and the health and lives of distant future people, require that we proceed with caution in this endeavor. Gregory Kavka aptly expresses this sentiment in a way that incorporates the biological, metaphysical, and moral dimensions I have discussed: "If our capacity to manipulate genes develops so that we determine the identity of our offspring in a biological as well as a social sense, this may influence—in ways that are not benign—our *collective consciousness*, including our conception of who we are and what our place is in the overall scheme of things."[49]

2

GENETIC INFORMATION, OBLIGATION, AND THE PREVENTION OF LIVES

Questions about whether we can or should prevent early- or late-onset diseases in people who do not yet exist and those who already exist presuppose that we have a certain degree of control over the incidence of disease. This raises further questions about responsibility and obligation. In general, one is causally responsible for a state of affairs just in case one causes it to obtain or has control over the events that cause it to obtain. One is morally responsible for a state of affairs just in case one is causally responsible for it and is capable of conforming one's behavior to social norms concerning how one ought to act in certain circumstances. Moral responsibility presupposes casual responsibility, which in turn presupposes casual control over states of affairs.[1] But to be morally responsible for a state of affairs, one must have an obligation to cause it to obtain or prevent it from obtaining. In this chapter, I will discuss the extent of our obligation to prevent the states of affairs identified with genetically caused diseases and our responsibility for them if we allow them to obtain.

There are metaphysical and epistemic aspects of causal control and responsibility. The first aspect says that a person has causal control over, and thus can be responsible for, a state of affairs if he is not forced to act by external factors and is not compelled to act by internal factors such as irresistible impulses. In addition, the state of affairs must be causally sensitive to the person's action in the sense that, if the person had done otherwise, then it would not have obtained. The second aspect says that a person has causal control over, and thus can be responsible for, a state of affairs if he is capable of having appropriate beliefs about the circumstances in which he acts and the foreseeable consequences of what he does or fails to do.[2] Again, moral responsibility for a state of affairs presupposes a moral obligation to prevent it or bring it about. Although both metaphysical and epistemic as-

pects of causal control are necessary and sufficient for a person to be responsible for a state of affairs, it is the epistemic aspect of control, obligation, and responsibility with which I will be concerned here. I will analyze what our knowledge of genetic information and the risk of harm it
entails in the form of disease implies about our obligation to prevent this
harm and our responsibility for it if we allow it to occur. More specifically,
I will consider whether those who have a genetic mutation with a high
probability of causing a disease have an obligation to inform other family
members who also are at risk of developing that disease. More controversially, I will consider whether potential parents have an obligation to prevent the existence of people they can foresee who would experience severe pain and suffering and whether parents can be responsible if they
bring them into existence.

Our knowledge of genetic information and responsibility for how it is
used or acted upon assume the ability to test cells of embryos, fetuses, and
persons for genetic abnormalities that predispose them to or directly cause
disease. I will focus mainly on two forms of genetic testing: presymptomatic
testing of persons for mutations causing adult-onset disorders such as Huntington's disease and breast cancer; and testing of preimplantation embryos
for mutations causing early-onset disorders such as Duchenne muscular dystrophy, Tay-Sachs disease, and Lesch-Nyhan syndrome, as well as for late-
onset disorders like Huntington's and Alzheimer's diseases. The first type of
testing is done on actual people, while the second type is done on early-
stage human organisms that may or may not develop into people.

In the first instance, I will argue that individuals who know that they
have the mutant allele of the BRCA1, BRCA2, or Huntington's genes, for
example, have an obligation to inform their children or siblings that they
too are at risk of developing the disease and allow them to choose whether
to be tested for the presence of the mutation. In the second instance, I will
argue that, if parents know that there is a high probability of transmitting a
mutation that would cause a severe disease in a child and have access to affordable reproductive technology, then they are obligated to have embryonic or fetal cells from which the child would develop tested for the mutation. If the cells do contain the mutation and the disease it causes is severe,
then the parents have an obligation to terminate the further development
of the affected embryo or early-stage fetus. This argument is motivated by
two related moral principles. As a matter of nonmaleficence, we have an
obligation to prevent pain and suffering in the people we bring into existence. As a matter of justice, we have an obligation not to cause people to

exist with cognitive, physical, or emotional disorders and disabilities that would severely limit their opportunities for achieving a minimally decent level of lifetime well-being.

In assessing the risk of inheriting or passing on a disease with a genetic cause, it is not only the probability of the gene causing the disease that matters, but also the severity of the disease and the number of people who may be affected by it. These three factors indicate that the moral obligation to be tested for the presence of a genetic mutation and to share this information with others is a matter of degree. When the probability that a mutation will cause a disease is high, when the symptoms of the disease are severe, and when many people may be affected, the obligation will be stronger. When the probability that a mutation will cause the disease is low, when the symptoms are mild to moderate, and when few people may be affected, the obligation will be weaker. Indeed, there may be no obligation in such a case. The particular obligation to acquire and share genetic information about diseases is grounded in the principle that one has a more general moral obligation to prevent harm from befalling others, provided that one does not do more than minimal harm to oneself as a result.

First, I examine the content of obligation and responsibility, of what we are obligated to do and our responsibility for what we do with the information derived from genetic testing for early- and late-onset diseases. I then explain why we are not obligated to bring any people into existence, as well as what our obligations are to the people we do bring into existence in terms of the moral principles of nonmaleficence and justice. This will require addressing objections that the position for which I am arguing would lead to perfectionism and discrimination against people with disabilities. Construing "worse off" in terms of the presence or absence of disease and disability, I go on to discuss the egalitarian principle holding that a smaller benefit to the worse off (sick) morally outweighs a greater benefit to the better off (healthy). I point out that how we interpret this principle will be sensitive to the number of people who stand to benefit or be harmed and the magnitude of benefit or harm at issue. Finally, in summarizing my arguments, I reiterate that they are motivated and supported by the moral importance of preventing harm to people.

Genetic Testing, Screening, and Information

Genetic testing must be distinguished from genetic screening.[3] The first refers to testing individuals who are known to be at increased risk of having a genetic disorder with a familial mode of inheritance. The second refers to testing members of a particular population for a disorder for which there may be no family history or other evidence of its presence. Let us consider some cases involving information derived from genetic testing and genetic screening that illustrate the importance of the epistemic aspect of obligation and responsibility in preventing harm to others.

Suppose that a man with a family history of Huntington's disease suspects that he may have the disease after developing jerky bodily movements. Huntington's is caused by the insertion of multiple in-frame repeats of the CAG codon (encoding glutamine) into the IT15 gene.[4] A single mutated allele of the Huntington's gene is sufficient to cause this ultimately fatal disease. It is characterized by irreversible motor and mental deterioration and considerable pain for the patient, as well as considerable suffering for both the patient and his family. As a dominant disorder, persons with Huntington's can transmit the gene and the disease to half of their offspring. For those who have the gene, there is virtual 100% penetrance, meaning that they most certainly will develop the disease. Unfortunately, by the time symptoms appear, they already may have had children and unwittingly transmitted the gene and the disease to them. However, there is now a test that can determine whether an individual has the gene before symptoms appear. Given the predictive power of genetic testing, would the man in our example have an obligation to be tested despite the fact that there is no effective treatment for the disease?

To the extent that the man had been diagnosed with Huntington's and understood the gene's degree of penetrance and the progression of the disease, he would be obligated to be tested and inform any of his children of the result. Testing would confirm the diagnosis and would clarify the 50% risk for his children. From the children's perspective, the difference between a 0% and 25% risk, or between a 25% and 50% risk, is significant. The difference in risk can influence important life choices. The father would be obligated to be tested and to share the information because it would prevent harm in two respects. First, by having precise information about the risk, his children may decide not to have children of their own and thus ensure that they would not transmit the gene and the disease to any additional people. Second, the information would better serve his children's prudential interests, enabling them to plan other aspects of their lives accordingly. The testing and sharing of genetic information could minimize lost opportunities and maximize opportunities for achievement within a shorter life span with severely disabling middle and final stages.

The point here is to offer those at risk of a severe disease the opportunity to make informed choices, given how they assess the risk of having the gene and the fact that its virtual 100% penetrance means that they would develop the disease if they had the gene.

Informing a son that he had a 50% risk of having Huntington's, which could begin affecting him in the prime middle stage of his life, could allow him to choose to be tested, and depending on the result, enable him to plan the remainder of his life in his best interests. For example, knowing that he had the gene and would have the disease, he might give up plans for a career in medicine, which has a long period of apprenticeship, and opt instead for a career that would enable him to have and realize opportunities for achievement earlier in life. Not being tested in the face of evidence of the disease in his family may be imprudent for such a person. Failing to plan one's life given the knowledge that one will have a severe disease in middle age may entail foregone opportunities earlier in life and foreclosed opportunities later in life. But insofar as prudential considerations alone are at stake, a single individual who believes or knows that he is at a significant risk of contracting a genetic disease would not be morally obligated to undergo genetic testing. There would be no others to whom he might be obligated.

To be sure, the obligation to inform siblings or children that one has a monogenic disease may be mitigated by the potential for psychological harm to oneself. This could be in the form of emotions such as anxiety, guilt, shame, fear of blame from family members, or fear of discrimination by prospective insurers or employers. Still, this psychological harm to oneself would have to be weighed against the magnitude of the potential harm in not informing others of their risk. If the number of people who may be affected is significant, then the potential for physical and psychological harm to others who remain ignorant may outweigh the potential for psychological harm to the individual deciding whether to be tested and to share the information with them. In the case of Huntington's in a parent with many children, there would be moral grounds for saying that an individual with symptoms and a diagnosis would be obligated to prevent the greater harm by being tested and informing others of his disease and of the fact that they too may be at high risk of developing it.

Consider now a case illustrating the justification for preconception genetic screening within certain communities. There is a much higher incidence of Tay-Sachs disease within the Ashkenazi Jewish community living in different countries. Individuals affected by this disease usually appear quite normal at birth. But in the first year of life their nervous systems degenerate, and they usually die by the time they reach three or four. Tay-Sachs is an autosomal recessive disorder, where the affected child inherits one mutant allele from each parent, both of whom are only carriers of the trait and do not have the disease. Because of screening programs, there has been a significant reduction in the incidence of Tay-Sachs. This is due to carriers avoiding marriage in Orthodox Jewish communities, to carrier couples undergoing prenatal testing and terminating affected embryos or fetuses, to the use of donor gametes, and to adoption. Adolescents constitute a large percentage of those screened, which has prudential and moral implications. Doing so at an early age allows them ample time to plan their future marriages and family lives in accord with their values and to prevent the harm that would result from having a child with Tay-Sachs disease.

One concern with genetic screening programs is that they cause undue anxiety and thus can harm those who participate in them. But the following account of a pilot study of carrier screening for CF through a blood test among 341 secondary school students in Montreal shows that anxiety does not always result from screening.

Students over fifteen years of age were invited to attend an information session that covered basic genetic information about cystic fibrosis, and the implications of being a carrier. A week later, the students were invited to participate in the screening program. Three of the nine carriers identified had experienced some anxiety on receiving the positive result and one person had great anxiety. This anxiety had dissipated in all the participants by the time of the follow-up interview.[5]

The most compelling case for genetic screening is for treatable disorders. What comes to mind most readily is the autosomal recessive disorder PKU, which consists in the body's inability to metabolize phenylalanine, an amino acid, and which can lead to severe mental retardation in affected children. This can be avoided by a diet low in phenylalanine. Neonatal screening provides parents of affected children with the knowledge they need to administer the appropriate diet and is thus a relatively straightforward method of preventing disability and harm.

Considerations of harm also apply to testing for mutations in the BRCA1 and BRCA2 genes, which entail a significantly high risk of breast (up to 85 %) and ovarian (up to 60 %) cancer over a lifetime. As a dominant disorder, children of people with these mutations have a 50% chance of also carrying the mutations. Suppose that a woman in her forties whose

mother and grandmother both died from breast cancer at a relatively early
age suspects that she has the mutation. If she has no sisters or children, then
she would have no obligation to be tested and confirm whether in fact she
has it. There would not be any concern about harm to others. The only
concern would be a prudential one pertaining to her own interests, and
whether being tested and having the knowledge of whether she did or did
not have the mutation served these interests. This might include more frequent mammography starting at an earlier age. However, if she has been
diagnosed with breast cancer and has sisters or daughters, then she would
be obligated to be tested to confirm that it is caused by the BRCA genetic
mutation and to clarify the risk to them. The obligation would be
grounded in the combination of the high risk of breast cancer entailed by
the mutation and the high risk of morbidity and mortality for those who
have it, multiplied by the number of women in the family. With respect to
the penetrance of the gene, though, the obligation here would not be as
strong as it would be in the case of Huntington's. This is because of the difference between the 50–85% and 100% probabilities of having the diseases, given the genes in question.

One might claim that, insofar as the woman's daughters know that there
is a family history of breast cancer, the parent would not have an obligation
to be tested or inform them of the result. On the basis of this knowledge,
the daughters would be left to decide which course of action to take. But
suppose the mother knows that she has the gene. In that case, the fact that
the risk of having the disease entailed by this particular genetic mutation is
significantly higher than the risk entailed by other mutations implicated in
breast cancer, that children of people with the BRCA1 and BRCA2 mutations have a 50% chance of also carrying them, and that there is a high
probability of premature mortality with the disease, would obligate her to
tell them that she has it.

Still, one could insist that the woman has a right *not* to know whether
she has the mutation and therefore the right not to be tested.[6] The right to
choose whether or not to undergo genetic testing, which her daughters
would have on being told by her that they are at risk, is something that she
should have as well. Why should she waive her right not to be tested just so
that others might have this right to choose? This seems unfair to her. Nevertheless, most rights are not absolute but prima facie, in the sense that
they can be overridden by other considerations. One of these considerations is the probability of harm to others in exercising a right. When others
may be affected, their collective right not to experience avoidable physical

and psychological harm may override the particular right of the woman
not to experience any psychological harm in knowing that she has the
mutation.

Ordinarily, moral obligations to family members are generated not by
genetics but intimacy, the emotional bonds one develops in virtue of sharing needs and interests with others.[7] In the case of living-organ donation
for transplantation, for instance, one may have a greater moral obligation to
donate bone marrow to a sibling, parent, or child with leukemia than a
stranger because of the special relationship between them. This obligation
must be weighed against the degree of risk involved, which is why there
would be less of an obligation to donate a kidney or part of a liver than
bone marrow. But there is an important disanalogy between living-organ
donation and genetic testing for breast cancer. In the first case, there is an
obligation to aid someone who has the symptoms of a life-threatening
condition. In the second case, there is an obligation to get information
about a condition for which one is presymptomatic and which is more
treatable. Moreover, the risk of harm in living-organ donation is more significant than the risk in genetic testing because it is both physical and psychological, not just psychological. In the type of breast cancer at issue, it is
not intimacy but genetics, and the moral obligation to prevent avoidable
harm, which ground the obligation to be tested and inform one's children
of the result.

With the information that their mother's breast cancer was caused by
the mutation in the BRCA gene, daughters could choose whether or not
to be tested on both moral and prudential grounds. A daughter who is
contemplating having a child might choose to be tested, in which case a
positive result might cause her to change her reproductive plans on the
moral ground that she should prevent potential harm to a potential daughter by transmitting the mutation and perhaps the disease to her. This
daughter also might alter her life or career plans on prudential grounds,
depending on whether she undergoes genetic testing and how she assesses
the information about her risk of having breast cancer. She might opt for a
prophylactic mastectomy and oopherectomy to eliminate or reduce the
risk and avoid having the disease.[8] All of this would hinge on her assessment of the 50–85% probability of developing the disease given the genetic mutation, a probability high enough to obligate her mother to inform her of this risk.

Ideally, a woman with the BRCA1 or BRCA2 mutation could avail
herself of gene therapy to correct the mutation and thereby eliminate her

risk of breast cancer without surgery. But as I will explain in more detail in chapter 3, gene therapy still is not feasible for treating most genetic diseases. If she is considering having children, then she could take fertility drugs to induce superovulation and thereby produce multiple eggs that could be fertilized by a man's sperm in vitro to form multiple embryos. These embryos could be tested for the presence of the mutation, after which a genetically normal embryo could be selected for implantation in the uterus and development into a birth. If it were difficult to determine whether the embryos had or lacked the mutation, or if the mode of inheritance were such that all female embryos contained the mutation, then she could select only male embryos for implantation. (While males can develop breast cancer, the incidence of it is much lower than in women. Still, in a family with the BRCA mutation, any male member should be told that he could pass on the mutation to a daughter once he reaches reproductive age.)

Selecting embryos on the basis of sex may seem morally objectionable. But selecting male over female embryos to prevent a disease like breast cancer caused by either BRCA mutation would be morally justifiable. For similar reasons, a parent or parents could justifiably select a female over a male embryo in order to avoid such sex-linked disorders as Duchenne muscular dystrophy (DMD) and hemophilia, which only affect males.[9] The rationale for selection in all of these cases would be to prevent disease in and therefore harm to people.

If a person knew that she had a genetic mutation and a high risk of having a genetic disease, but the symptoms associated with the disease were only mild to moderate, then she would have a weaker obligation to inform other family members of the risk. For the likelihood of harm to them would be relatively low. Indeed, some might insist that in such cases the person would have an obligation *not* to inform others, owing to the unnecessary psychological harm that anxiety and other emotions might occasion in them. The same could be said about diseases with a genetic component that affect people near the end of a normal life span, such as the more common form of Alzheimer's disease. But in the small class of cases of early-onset Alzheimer's, as well as early-onset Parkinson's, both of which have a strong genetic component, there would be compelling reasons for individuals at high risk to be tested and to inform family members of their genetic status.[10] This is because of the severity of the symptoms of the disease and the period of time between the onset of symptoms and death, factors that can have a significant impact on the lives of those who have the disease and on family members or others who care for them. Knowing in advance that one's cognitive and physical condition will begin to deteriorate relatively early in life can allow a person and others caring for him to plan for the time when he and they will be most adversely affected by the disease. The moral and prudential reasons for being tested and sharing genetic information in cases of earlier-onset Alzheimer's and Parkinson's parallel the reasons for doing the same in cases of Huntington's disease.

There is also a legal aspect to the obligation to obtain and act on genetic information indicating a significant probability of transmitting a genetic disease to a child. To the extent that embryonic cells can be tested for genetic mutations that cause severe disease and disability, parents who had affordable access to the technology would have an obligation to test IVF preimplantation embryos produced from their egg and sperm if they knew that they were carriers of a genetic mutation. On the basis of the information derived from the testing, parents could decide to terminate an embryo with the mutation or allow it to implant and develop into a person. Yet if they decided to allow it to develop with the knowledge that there was a high probability of it causing a severe disease, then they would be responsible for their child's condition. A child caused to exist in such a condition could file a tort of wrongful life against his parents, or against an obstetrician or genetic counselor for misinforming his parents about the risk. If the child was not mature enough or was cognitively or physically unable to do so, then an adult could file a tort on the child's behalf. The justification for the tort would be that the parent, obstetrician, or genetic counselor was legally responsible for the child's condition and accordingly owe it compensation for acting negligently or recklessly in ignoring the risk, given the parents' carrier status and the probability of genetic transmission.[11]

Suppose that a boy has DMD, a recessive sex-linked disorder traceable to an X-chromosome mutation that adversely affects the function of the dystrophin protein. Half of all boys who are born to women with the mutation are affected. The defect causes muscles to begin weakening around age three and subsequent respiratory failure, giving the males afflicted with the disease an average life expectancy of sixteen to twenty years. If the boy's mother knew that she was a carrier of the mutation and that there was a 50% chance of transmitting it to a male child, was able to test embryonic cells in vitro or fetal cells in utero through amniocentesis, and still allowed the embryo and fetus to develop into the child, then the son could claim that his mother (or both parents) acted in reckless disregard of his welfare.

He could claim that she, or they, harmed him and owe him compensation for causing him to exist in a condition that defeats his interest in having a life that is not so severely restricted by the disease.[12] The child has to live with a substantial burden that was imposed on, not consented to by, him. His parents would be morally and legally responsible for his condition on the ground that they knew, or were reasonably expected to have known, that there was a significant risk of having a child with a condition that would harm him. Alternatively, his parents might have produced multiple IVF embryos and selected all female or only female noncarrier embryos for implantation and further development. This would be a morally justifiable example of sex-selection as a method of preventing significant harm to the people we bring into existence, given that DMD is a sex-linked disorder affecting only males.

Some might claim that parents have an obligation to ensure that their children have the best opportunities in life for achievement and well-being. Suppose that, in addition to the sex of the embryos, the likely sexual orientation of embryos could be ascertained either in vitro or in utero. Because women and homosexual males have been harmed by discrimination, it would seem that parents have an obligation to prevent harming their children by selecting only heterosexual males. But it is doubtful that any burdens resulting from being female or a homosexual male would be substantial enough to say that parents harm these individuals by bringing them into existence. Moreover, I have characterized disease and disability as conditions that harm the people we procreate, and neither sex nor sexual orientation is a disease or disability. Also, parents have an obligation only to ensure that their children have adequate opportunities for achievement, not the best possible opportunities. Parental obligation to children is grounded in nonmaleficence and beneficence, not perfectionism. Hence, while it may be permissible for parents to select the sex or sexual orientation of their child, they would not be required to do so.

All of the genetic diseases I have discussed thus far involve a combination of a high risk of transmission from parent(s) to offspring and a high risk of morbidity and premature mortality. Moreover, all of these diseases are monogenic disorders caused by a mutation in a single gene coding for some crucial protein regulating cell function. One multifactorial disease that complicates the issue of parental obligation is schizophrenia. Like bipolar disorder and major depression, schizophrenia tends to run in families. Parents with this mental disorder have about a 10% chance of transmitting it to their children. As already noted, this disease is caused by mutations in several genes adversely affecting the neurotransmitter dopamine. The mutations may result from interactions between genes and events in the uterine environment during gestation, or events in physical and social environments after birth. Symptoms usually appear in adolescence or early adulthood. Schizophrenia is treatable, but it requires considerable pharmacological intervention and social support to keep it under control. If social support is inadequate or absent, then the disease may become more severe and result in more suffering or even death (through suicide) for the person afflicted with it.

Consider a forty-year-old single woman who carries the mutations and has no extended family. If she is cognitively able to know that there is a risk of transmitting the mutations implicated in schizophrenia to her child, and that the requisite social support would be lacking, then there may be a reason for not having such a child. If the parent had this knowledge but decided to have a child who developed the condition, then arguably she would be responsible for that condition. Even so, the fact that social and biological factors beyond the parent's control play a crucial role in the etiology of schizophrenia, and that the risk of transmitting the disorder to children is only about 10 %, would constitute mitigating conditions, making the parent I have described at most only partly responsible for her child's disease. Unlike the previous cases I examined, here the rationale for not having a child would be tied not only to genetics but also to other biological and social factors that could affect the severity of the disease and be harmful to the person who has it. This example also illustrates that it is not only the risk of transmitting a genetic mutation that matters in assessing parental responsibility for conditions that affect their children, but also the magnitude of the harm their children may have to experience. And this may be due to nongenetic as well as to genetic factors.

Thus far, I have been discussing how our knowledge of the risk of transmitting genetic mutations to offspring and of the likely harm this risk entails bears on our obligation to prevent harm. I also have been discussing our responsibility for the harm these mutations cause if we bring people into existence who would have them. But more needs to be said about the philosophical basis of harm.

Moral Asymmetry and Harm

The principles of nonmaleficence and justice ground the following asymmetry thesis. We do not have a moral obligation to bring people into existence with good lives. But we do have an obligation to prevent the existence of people who would experience so much pain and suffering that their lives would not be worth living for them on the whole.[13] This thesis rests on two additional moral principles and the ontological distinction between human organisms (as embryos and early-stage, presentient, fetuses) and persons.

The person-affecting principle says that a person is benefited or harmed when her interests in what happens to her are satisfied or defeated.[14] Once a person exists, she has an interest in not experiencing pain and suffering and can be harmed if she experiences these over the course of her lifetime. The impersonal comparative principle says that, other things being equal, it is worse to cause a person to exist if it would be possible to cause a different, better-off, person to exist instead.[15] It involves a comparison, not between the existence and nonexistence of one person, but rather between two distinct lives of two distinct people. On the impersonal comparative principle, we evaluate two potential lives of two potential people who do not yet exist, while on the person-affecting principle we evaluate the life of one person who already exists. Yet we can appeal to both principles to support the claim that we are morally obligated to prevent the existence of people who would have lives that on balance would not be worth living. We prevent harm to the individuals we cause to exist by satisfying their interest in not having to experience severe pain and suffering, and we avoid adding to the total amount of suffering in the world.[16] This assumes that the diseases causing the pain and suffering that make life not worth living cannot be treated adequately.

In chapter 1, I argued that "person" is a psychological concept and that "human organism" and "human being" are biological concepts. These psychological and biological concepts are ontologically distinct because they involve distinct essential properties. I also argued that an embryo or presentient fetus is a potential person, not in the sense that it becomes a person, but only in the sense that it has the potential to develop the biological structures and functions necessary to generate and sustain consciousness and other forms of mental life definitive of personhood and personal identity through time. It is important to bear these points in mind for the discussion in the remainder of this chapter.

On the plausible assumptions that only beings with interests can be harmed by the defeat of these interests, that having interests presupposes sentience or the capacity for mentality, and that only persons and late-stage fetuses (and some nonhuman animals) have this capacity, it is morally permissible to terminate a human organism at an early stage of development. The termination affects no one who has interests. A person may be harmed by the later effects of parents' behavior during earlier fetal gestation, as in fetal alcohol syndrome. But it is the person who would be harmed by the defeat of his interests, not the early-stage fetus. One might object that having interests does not presuppose sentience. For example, it can be said that future people, who do not yet exist and therefore are not sentient, have an interest in not living in a polluted environment. But the core concept at issue here is harm, and harm most plausibly consists in the defeat of particular interests of identifiable persons who can experience the defeat of these interests. Terminating the development of a human organism at the embryonic stage does not kill a person but only prevents a person from coming into existence. There is no one who could be harmed because there is no identifiable individual with particular interests who exists at the time of the termination.

Similar reasoning underwrites the claims that we do not have a moral obligation to bring people into existence, and that bringing someone into existence by itself does not benefit her, however good her life may be. An obligation implies that some people have an imperative to fulfill a claim that others have made on them. Nonexistent, potential people cannot make claims of any sort, including the claim to be brought into existence. Furthermore, we cannot say that causing a person to exist makes her better off than she otherwise would have been, since otherwise *she* would not have existed. Put another way, there is no one *who* is caused to exist; rather, we make it the case *that* someone exists. Derek Parfit argues that the relation between existence and nonexistence does not meet the "full comparative requirement," which says that we benefit a person only if we do what will be better for him.[17] While causing a person to exist may be good for him, it cannot make him better off. For nonexistence is morally neutral, neither good nor bad, and therefore it cannot coherently be compared with existence. In causing a person to exist, we cannot make him better off than he was before he existed, on pain of contradiction. So we cannot benefit persons by bringing them into existence. We benefit people by satisfying their interests, and it is difficult to understand how a nonexistent person could have an interest in being caused to exist.

This is not to say, however, that we cannot coherently compare longer and shorter lives of the same person and determine whether he would be better or worse off in one life or the other. If a sudden premature death prevents an otherwise healthy person from experiencing goods in the future, then a shorter life makes him worse off than he would have been had he continued to live. He would have benefited from a longer life. In contrast, if a person with a severe chronic disease can only look forward to more pain and suffering without any compensating goods, then a shorter life with an earlier death would make him better off. He would be harmed by a longer life. What makes these comparisons between shorter and longer lives of a person coherent is the fact that they are future possible lives of one actual person who already exists and has interests.

Although it is morally neutral to cause a person to exist with a life that on balance is good, it is morally wrong to bring a person into existence with a life that on balance is so bad as not to be worth living. For in this case there *is* someone who actually experiences pain and suffering and thus is harmed by being caused to exist with such a condition. And because it is not morally neutral but wrong, we have a moral obligation to prevent the existence of a person who would have such a life. Yet there is an air of paradox about the idea that, while we do not benefit people by bringing them into existence with lives that are good on the whole, we harm people by bringing them into existence with lives that are bad on the whole.[18] Presumably, we harm a person by causing her to exist with a disease or disability because we make her worse off than she otherwise would have been. But if she did not exist with the disease or disability, then she would not have existed at all. Insofar as a person's life is worth living on the whole, being brought into existence with a cognitive or physical disability cannot be worse for her, because if we terminated the development of the embryo or fetus carrying the genetic mutation causing the disability, then she would not exist. This is a variant of what Parfit has called the "non-identity problem."[19]

Jonathan Glover has devised a strategy to sidestep this problem. Instead of trying to draw a comparison between the existence and nonexistence of one identifiable individual, Glover maintains that the relevant comparison is an impersonal one between two distinct lives of two distinct people. It is impersonal in the sense that "harm can be done even though identifiable people are no worse off than they otherwise would have been."[20] Glover uses the following example to illustrate the impersonal comparative principle. Imagine that a factory emits a chemical that causes babies to be born blind. According to the nonidentity problem, they are not made worse off than they otherwise would have been, since their lives are worth living and otherwise they would not have existed. Yet "what we should say here is, not that the pollution made the blind children worse off than they would have been otherwise, but instead that their condition is worse than the condition of the other children who would have been born in the absence of the pollution."[21] The choice is between bringing some or other people into existence, while retaining the same number of people who will exist.[22] Both person-affecting and impersonal harm principles give us reasons to bring a healthy child into existence rather than a diseased or disabled one. We prevent actual people from experiencing pain and suffering and thereby avoid defeating their interest in having healthy lives, and we avoid adding to the total amount of suffering in the world. Because these principles are grounded in nonmaleficence, beneficence, and justice rather than perfectionism, we are obligated only to ensure that the people we cause to exist have a minimally decent life, not the best possible life. Thus we avoid the negative utilitarian idea that no one should reproduce, since life is the condition for any harm that is experienced.

Against the position I am defending, Joel Feinberg holds that the argument for wrongful life is not based on comparing the actual condition of an individual with a counterfactual condition. Rather, a person wrongs another by bringing him into existence if his existence is such that never having existed would have been preferable.[23] But the coherence of the preferability claim relies on the impersonal comparative principle, where the comparison is between two distinct lives of two distinct people. One who exists cannot prefer a state in which one never existed. This is different from the claim that one can prefer ceasing to exist to continuing to exist, where the comparison is between longer and shorter lives of someone who already exists. In the DMD case, the "wrong" of wrongful life is explained by the impersonal comparative principle because it says that, given a choice between a better or worse state of affairs, we should choose the better. To the extent that the parents had a choice not to bring the DMD child into existence but a different or no child instead, they brought about a worse state of affairs. In addition, they harm the child by defeating his interest in not having to experience pain and suffering *once* he exists. The impersonal comparative principle explains the wrongfulness of the act of bringing the child into existence, and the person-affecting principle explains the harmfulness of the act. We need both principles to explain and justify what parents are obligated to do and what they are responsible for in causing people to exist.

David Heyd maintains that the interests of future, potential, persons cannot meaningfully be referenced to the choices and actions that bring them into existence.[24] This seems to be supported by my argument that potential persons do not become actual persons, in the sense of being identical to them. Yet assuming that any person who exists has an interest in avoiding pain and suffering, and that parents can foresee that a genetic mutation in tested embryonic or fetal cells entails a high probability of a condition with pain and suffering, the interests of the child can be referenced to the parents' choice and action. In choosing to bring a child into existence, the parents reasonably can be expected to have some understanding of the interests the child would have once he exists. Granted, there is no relation of identity between the embryo or early-stage fetus and the child. Nevertheless, there is enough of a causal thread running through the biological process of development from earlier to later stages to link the future interests of children not to have severe disease with the past reproductive choices of parents that resulted in the existence of these children. Parental obligation and responsibility transfer from their earlier reproductive choices to the later physical and psychological conditions of their children who exist because of these choices.

In some genetically caused diseases, severe cognitive and physical symptoms may make people's lives so painful and restrictive that they are not worth living for them. By definition, these lives fall outside the domain of the nonidentity problem. When we can predict that a disease would involve so much pain and suffering that the life of any person who had it would not be worth living on the whole, we are morally required to prevent the existence of people who would have the disease. Alternatively, if we do cause people to exist with severe diseases, then we are morally required either to cure them or alleviate their symptoms, insofar as we are able. The first scenario that I have described pertains to the impersonal sense of harm and potential persons, while the second pertains to the personal sense of harm and actual persons. One suggests genetic testing and selective termination of affected embryos as the appropriate action to prevent harm. The other suggests that the appropriate action to compensate for or prevent further harm is gene therapy or some other form of genetic intervention. A person is a further stage of development of the same human organism of which an embryo is an earlier stage. In this respect, embryos are potential persons, and failing to terminate a genetically defective embryo can cause harm by allowing it to develop into a severely diseased person and defeating its interest in having a healthy life.[25] Gene therapy has not been effective in treating most diseases. Because of this, it seems that genetic testing and selective termination of genetically defective embryos is the only medically effective and morally defensible way to prevent substantial harm to the people who come into existence. This suggests that some lives are not worth living and accordingly should be prevented, an idea that needs to be defended.

Should Some Lives Be Prevented?

Biotechnology enables, or will enable, us to test embryonic cells for genetic mutations that cause severe early-onset disorders, such as Tay-Sachs, Hurler syndrome, Lesch-Nyhan syndrome, and Canavan disease. These disorders have a similar profile. Hurler involves disruption of cognitive development in early childhood and usually death by age ten. Lesch-Nyhan causes both mental retardation and compulsive self-mutilation in boys. The irresistible urge to chew their lips and the tips of their fingers, in addition to experiencing the sort of extreme pain often associated with gout, obviously harms them. Even if they were unable to feel pain, the self-mutilation obviously would involve significant harm to them. Canavan is a degenerative disease that strikes infants, leading to decay of the nervous system and early death.[26] We also can, or will be able to, test for monogenic late-onset disorders such as Huntington's and forms of Parkinson's and Alzheimer's. Yet, as I have pointed out, the general ineffectiveness of gene therapy and other treatments for these disorders indicate that the best course of action is to prevent them by selectively terminating the embryos with the genetic mutations that cause the disorders. This means preventing diseases by preventing the lives of the people who would have them. Such a practice could be morally justified on two grounds. Nonmaleficence requires that we not harm people by causing them to experience pain and suffering over the balance of their lives. Justice requires that we not deny people the same opportunities for achievement and a minimally decent life that are open to others who are healthy or who have only moderate to moderately severe diseases.[27]

Arguably, the justice requirement will apply only to a small number of people. For the idea of equal opportunity for a good life implies a certain number of years to undertake and complete projects, and most people with severe early-onset genetic diseases have relatively short lives. Perhaps this is not the case with CF, where people who have the disease often live for thirty years or more. But whether one judges that a life with CF should be allowed or prevented would have to be informed by the fact that its severity spans a broad spectrum, from male adult infertility to constant severe life-threatening infections. Considerations of justice matter. But what matters more than ensuring equal opportunity for achieving a decent life is preventing avoidable severe pain and suffering that people will experience once they exist. This is what makes lives not worth living on the whole. Indeed, it is often the pain and suffering associated with severely disabling diseases that preclude people from having the opportunities necessary to achieve a decent minimum level of lifetime well-being.

Testing embryonic cells for genetic abnormalities is most effective with preimplantation embryos. This can be done once they have reached the eight-cell stage of development. To produce extracorporeal embryos for this type of testing, a woman can take fertility drugs such as Clomid or Pergonal to induce superovulation and a number of eggs that can be recovered for fertilization with sperm. One advantage of producing multiple IVF embryos is that, if a genetic abnormality is detected in one of them, it can be terminated and another, normal, embryo can be selected instead for implantation and further development. Through this practice of embryo selection, parents can prevent the existence of a child who would have a severe disease or disability, thereby preventing any burdens the disease would have on a child.[28] It also could result in the gradual elimination of a mutation from a population that is at high risk for a certain genetic disease. The ability to produce multiple IVF embryos is a necessary biological presupposition of the impersonal harm principle, since at least two embryos must be available for a parent or parents to choose to bring a healthy child into existence instead of one who is diseased.

There is another attractive feature of this practice. As Robert Edwards explains, "identifying embryos with genetic abnormalities would offer an alternative to amniocentesis during the second trimester of pregnancy, and the 'abortion' *in vitro* of a defective preimplantation embryo . . . would be infinitely preferable to abortion *in vivo* at twenty weeks of pregnancy or thereabouts as the results of amniocentesis are obtained."[29] Testing IVF preimplantation embryos for genetic abnormalities would be preferable to testing fetal cells for the same by amniocentesis or chorionic villus sampling (CVS) because, unlike these invasive procedures, it would not be painful to a pregnant woman and would avoid certain medical risks. Specifically, placing a needle into the uterus to extract cells from either the amniotic fluid or embryonic membrane triggers a miscarriage in some pregnancies. CVS also may cause limb deformities in the fetus. In addition, moral status depends on the capacity for mental life, which requires the presence of certain structures and functions of the brain. Because embryos lack these structures and functions and thus the capacity for mental life, they have no moral status. So there would be nothing morally objectionable about selectively terminating embryos. Embryonic testing also would be helpful in detecting conditions resulting from chromosomal abnormalities that are not inherited, such as Down syndrome. Here too it would be morally preferable to amniocentesis and alpha-fetoprotein analysis, which detect abnormalities later during fetal gestation. If these conditions were deemed severe, then a decision to terminate once the results of these tests were obtained would be more morally controversial than with embryos, given that fetuses are at a further stage of biological development.

Terminating the development of one IVF embryo and transferring a different one for implantation would mean that the life of one potential person was not allowed to become actual, and the life of a different potential person was allowed to become actual instead. The same number of people would exist; but they would be different people. Still, whether one or a different person is allowed to exist does not matter morally. Rather, what matters morally is preventing pain and suffering that actual people will have to experience. To the extent that an embryo whose cells contain a disease-causing genetic mutation would result in prolonged severe pain and suffering in a person, we are morally required to prevent the disease by terminating the further development of that embryo.

What must be emphasized, though, is that the moral obligation to terminate embryos and thereby prevent certain people from coming into existence pertains only to people who would have *severe*, not simply moderate or moderately severe, diseases. Only severe diseases make people's lives not worth living on the whole. People with a physical disorder like CF are able to form life plans and undertake and complete projects within these plans in such a way as to achieve a decent minimum level of well-being in their lives, short as they may be. As I have noted, however, the spectrum of symptoms associated with the disease would have to be considered in assessing the question of well-being for people with CF. In addition, in moderate forms of cognitive disorders involving mental retardation, such as many cases of Down syndrome, people are able to have happy and fulfilling lives. Indeed, some might argue that because people with Down's have limited cognitive and emotional capacities, they do not experience the same fear and anxiety about success and failure in life that the rest of us do and consequently may very well be happier. Nevertheless, the fact that there is a fairly high incidence of congenital heart disease in people with Down syndrome complicates the issue of the quality of their lives and must be factored into any evaluation of them.

The baseline of what constitutes the level of physical, cognitive, and emotional functioning necessary for a life worth living should be set quite low. It is only in severe diseases, involving prolonged pain and suffering and restricted lives, that the functioning of persons would fall below the baseline of what counts as a life worth living. The early-onset diseases that I mentioned at the beginning of this section fall into this class, and therefore lives containing these diseases should be prevented. Because they would fall on or above the critical level, however, there would be no moral requirement to prevent the lives of people with Down's or CF. But there would be no moral requirement to bring them into existence either, given that existence itself is morally neutral and entails no obligation to cause people to exist with good lives.

With respect to the question of whether a life is worth living, it is important to distinguish early-onset genetic disorders of the sort I have been discussing, which affect people from birth or early childhood, from the late-onset genetic disorders of Huntington's, Parkinson's, and Alzheimer's. It is also important to distinguish disorders that affect people in mid-life from those that affect them much later in life. In Parkinson's and Alzheimer's, the earlier-onset and more severe forms of the diseases are correlated with a higher-penetrance genetic component. The severity and time of onset of symptoms affecting adults are crucial to assessing the value of their lives on the whole.

Life Stages

The onset of symptoms in people with Huntington's may range anywhere from age thirty to sixty, and these include progressive loss of muscle control and dementia. Generally, they die within fifteen years after onset. Prior to this time, they can have normal lives with comparatively high levels of cognitive and physical functioning for a considerable number of years. In trying to determine whether people's lives are worth living on the whole, we need to evaluate the quality of their lives in terms of all the stages in them. Quality of life is a function of the presence or absence of pain and suffering, and of the cognitive, emotional, and physical ability to have opportunities for achieving a decent minimum level of well-being. This includes the ability to form a life plan and to undertake and complete projects within such a plan. Obviously, this determination will be subjective to a certain degree. Nevertheless, by setting a low threshold of what counts as a decent minimum, there can be reasonable objective agreement about lives that are not worth living because they fall below the baseline. With late-onset diseases like Huntington's, though, the radical difference between cognitive and physical functioning before and after the onset of symptoms makes it difficult to assess overall quality of life for the people who have them. This is so even if we can agree on a general objective measure of quality at particular stages of life.

One way of measuring lifetime quality in severe adult-onset diseases is to weigh the level of normal cognitive, emotional, and physical functioning per year lived against the level of pain, suffering, and disability per year lived. From this we can roughly calculate an average level of well-being for the person's entire life.[30] On this model, the more severe the pain, suffering, and disability associated with the disease, the earlier the time of onset of symptoms, and the longer the period of time between onset and death, the stronger will be the reason for saying that the life is not worth living on the whole. Correspondingly, there will be a stronger reason for preventing such a life by terminating the embryo known to have the mutation that causes the disease, assuming that the risk of having the disease is known to be high. For a person who first experiences the degenerative physical and cognitive symptoms of Huntington's at age thirty-five and who dies at fifty, the severity of the pain and suffering he experiences in his last fifteen years may be bad enough to outweigh the good functioning he had in his first thirty-five years and average out to a level of lifetime well-being that falls below the decent minimum.

This assessment is influenced by the idea that we have a temporal bias toward the future. That is, we care more about what we will experience in the future between the present and the time of our death and less about what we experienced in the past.[31] Most of us prefer a life that matures to one that degenerates, a life that starts poorly and gets better to one that, traversing all the same points, starts well and then goes progressively downhill.[32] If this is plausible, then the subjective value of our last stage of life will be disproportionate to the value of other stages with respect to its overall quality. The suggestion is that a long and painful final stage of one's life may have enough disvalue for one to conclude that the life is not worth living on the whole. Accordingly, if a parent or parents can foresee such a life through the results of genetically testing the cells of an embryo, then they should terminate the embryo and prevent that life and the existence of the person who would have it.

Consider now a variant of the Huntington's example. Suppose that genetic testing of the embryo could predict that symptoms would not appear until age fifty. One still might claim that the pain and suffering caused by the disease in the last ten or fifteen years of the person's life would be so severe that they would outweigh even the good fifty years and thus make the person's average lifetime well-being fall below the decent minimum. Although it would be more difficult to sustain than in the earlier example, here too there may be a principled reason for terminating an embryo with the Huntington's gene. It may weigh the decision in favor of preventing the existence of a person who would have the disease over causing him to exist with a life that has a significantly long and wretched last stage. Moreover, suffering may begin much earlier than the time of onset of symptoms. Knowing that one is at risk of having the disease and the symptoms it entails can influence a negative perception of the quality of one's life even before symptoms begin to appear.

Alternatively, one could take the view that, with adult-onset diseases, the person with the disease would have the autonomy and responsibility to decide for himself whether to go on living beyond a certain point. This would shift the locus of decision-making and responsibility from the time before birth to the time between the onset of symptoms and death. It also would shift the responsibility for decision-making from parents or other parties to the person with the disease. This is consistent with the conviction that the value of a life is largely (though not entirely) determined subjectively by the person whose life it is. This value has much to do with the subjective notions of dignity and control over how the remainder of one's life should go. But these notions invite questions about voluntary euthanasia and assisted suicide, which involve legal issues that are beyond the scope of this essay. More important for present purposes, a parent or parents would be justified in preventing a potential person from existing if a late-onset disease the person would have involved severe mental and physical symptoms and a considerable period of time between the onset of symptoms and death. The asymmetry in our attitudes toward earlier and later periods of life and the person-affecting principle justify an objective determination that such a life on the whole would not be worth living for any person.

A mutation in the gene for beta-amyloid precursor protein (BAPP), when combined with apolipoprotein E (APOE), has been implicated as a cause of Alzheimer's disease. BAPP gathers in the spaces between nerve cells in the brain, while APOE transports cholesterol in the bloodstream and plays a role in cellular repair and regeneration. APOE comes in three alleles, e2, e3, and e4. Patients with the BAPP mutation and the e4 allele of

APOE develop Alzheimer's much earlier than those with the BAPP mutation and the e2 or e3 alleles.[33] The average age of onset is sixty-eight years
in people with two copies of APOE4, seventy-five years with one copy,
and eighty-four if they have no APOE4 genes. Suppose that a person with
one or two copies of the e2 or e3 allele develops the disease at eighty-five.
Given that she already has lived out a normal life span, that she probably
would die in a few years, and that she has had no major diseases or disabilities prior to this time, the balance of benefits to burdens in her life as a
whole would indicate that her life is worth living and that there would be
no reason to prevent it. But suppose that a person with two copies of
APOE4 has a much earlier-onset form of the disease than the average for
this type. If symptoms began at age forty or earlier and the individual's
condition deteriorated gradually over a period of twenty years, then one
could argue that the balance of burdens to benefits in that person's life as a
whole would make it not worth living and thus provide a reason for preventing a similar life in the future. It is not so much the final stage of the
disease that is harmful to those affected by it, since they may no longer be
aware that they have it. Rather, the harm consists more in the suffering associated with experiencing the gradual deterioration of one's mental and
physical functioning over a prolonged period.

Admittedly, these assessments are always probabilistic, as many people
with BAPP and APOE4 never develop Alzheimer's. Nevertheless, people
with two copies of APOE4 have about a 90% chance of eventually developing Alzheimer's. If genetic testing of embryos could be perfected to the
point where the difference between the e2, e3, and e4 alleles of APOE and
their probability of penetrance could be determined, and if the onset of
symptoms in people who develop from these embryos could be predicted
with a high degree of probability, then this could influence decision-making regarding which lives are allowed and which are prevented. What is
crucial is not just the nature and progression of the disease, but also *when* it
occurs.

Pain and suffering in the last stage of one's life must be weighed against
the positive experiences and achievements afforded by normal mental and
physical functioning in earlier stages. Fifty years of normal functioning
should be enough for a person to complete many of the projects in her life
plan and make for a life that is worth living. If we accept this, then the fact
that a person's normal functioning begins to deteriorate at age forty by itself is not enough to support the claim that on balance her life is not
worth living and that she should not have been brought into existence.

Some people have lives of just thirty years that are full of achievements and happiness. A shorter life can be very well worth living.³⁴ But if the pain and suffering in one's last years are severe, and if the number of these years is considerable, then this may weigh the decision in favor of preventing the person from coming into existence. The experience of severe pain and suffering, more so than what it implies about limited life opportunities, grounds the claims that a person's life on balance is not worth living and that it is morally wrong to cause a person to exist with such a life.

Still, we have to weigh the relative importance of cognitive and physical functioning for different people in evaluating quality of life. Consider the motor-neuron disease amyotrophic lateral sclerosis (ALS), where 5 to 10% of cases are familial. Mutations in the gene encoding the enzyme superoxide dismutase 1 (SOD1) account for 20% of cases of familial ALS.³⁵ The adult-onset form of the disease is autosomal dominant, with symptoms appearing around age forty or earlier. A juvenile-onset form is autosomal recessive. In all familial cases, penetrance is close to 100 %. Suppose that the presence of the mutation in the dominant type could be detected by testing embryonic cells, and that the disease could be prevented only by terminating affected embryos. One might think that there would be strong medical and moral reasons to terminate such embryos, given the degree of penetrance, the symptoms associated with the disease, the time of onset, and the likely duration of the disease. Yet the case of the brilliant theoretical physicist and cosmologist Stephen Hawking illustrates that a person can suffer from a severe physical disability as a symptom of ALS for well over the balance of his life and yet maintain a high level of cognitive functioning that makes life very much worth living for him. It would be difficult to adduce reasons for preventing his life, despite the fact that ALS is a severely disabling disease. On the other hand, for people who value physical functioning very highly (for example, athletes, or dancers), severe physical disability may lead them to judge that on balance their lives are not worth living, even if their normal cognitive functioning remains intact. Granted, people can adapt their preferences and life plans to adult-onset diseases and the limited opportunities they entail. But it is not so easy to do this if one has to endure constant pain and suffering over a long period of time, especially if symptoms appear early in one's life. Furthermore, in the case of someone like Hawking, if the pain associated with his physical disability were so severe that it adversely affected his cognitive functioning, specifically his ability to do physics, then we might ask whether he would believe that his well-being was at a level high enough for him to judge that his life was worth living.

These considerations suggest that how a late-onset disorder with variable symptoms and variable degrees of pain and suffering affects a person's quality of life will depend on the type of life plan the person already has chosen and cultivated before the onset of symptoms. A person's life in progress is shaped by the person's values and preferences. And these are not the sorts of things that can be predicted on the basis of genetic testing of the embryo or early-stage fetus from which the person develops. This point likely would not arise in a case where juvenile-onset ALS could be predicted on the basis of genetic testing of an embryo. Like other early-onset neurodegenerative diseases, there would be a decisive reason for preventing the life of the person who would have the disease.

In moderate and moderately severe diseases with a genetic cause, we could argue that we would not be *required* to prevent people from coming into existence with these diseases. We would be *permitted* to prevent them from existing if we knew or had good reason to believe that existing would entail some harm to them. But we would be required to prevent their existence if it entailed severe disease and considerable harm. In theoretical terms, an action is morally required if it is supported by a decisive reason for doing it. An action is morally permitted if there is no decisive reason for not doing it. What permits it is the absence of a reason sufficient to ground a prohibition against the action.[36] In practical terms, the difference between requirements and permissions is a function of the probability of having a disease and the magnitude of harm in the severity and duration of the symptoms associated with the disease. If the probability of having a disease and the magnitude of harm are high, then there is a decisive reason for preventing the life of a person who would have a severe disease. Thus we are morally required to prevent the existence of such a person, or morally prohibited from causing the person to exist. If the probability of having a disease and the magnitude of harm are low, then there is no decisive reason for preventing the life of a person who would have a moderate disease. Thus we are morally permitted either to prevent the person from existing or to cause him to exist. The harm would not be so great as to constitute a decisive reason for preventing the life of a person who would experience moderate pain and suffering. In no case, though, would we be morally required to cause people to exist with healthy lives, since existence is morally neutral and causing them to exist does not make them better off.

Some might be disturbed by the idea of making decisions about embryos on the basis of the probability of developing a disease, given a genetic mutation. Even with a family history of a genetic disorder, the gene in question may not express itself in every member of the family. Moreover, even if one inherits a gene entailing a fairly high probability of disease, diet and lifestyle may prevent it. This is supported by Holtzman and Marteau's point, noted in chapter 1, about the complexity of genetics in common diseases in the light of the incomplete penetrance of genotype in these diseases.

But it is not just the probability of having a disease, given a particular genetic mutation, that constitutes a reason for preventing a life. In addition, the magnitude of harm entailed by the symptoms of the disease must be considered, as well as the time of onset of symptoms and the time between onset and death. Earlier, I specified these three criteria in assessing whether a life is worth living on the whole. The probability of penetrance is a further criterion that must be specified in judging whether a life should be prevented. So, assuming that a defective gene can be detected at the embryonic stage, four criteria will figure in determining whether a life should be prevented by terminating the embryo with the gene: (1) the probability of genotypic penetrance, or how likely it is that a genetic mutation will cause a disease; (2) the severity of the disease in its symptoms; (3) the time of onset of symptoms; and (4) the time between onset and death.

It will be helpful to think of the obligation to prevent a life as a matter of degree, falling along a continuum between requirements and permissions. The obligation will be stronger or weaker depending on the strength of the reasons for or against preventing a life, which in turn will depend on how a particular instance of disease fits the four criteria specified above. In a disease with a high probability of penetrance and with an onset of severe symptoms early in life, there is a strong, decisive, reason and obligation to prevent the life of the person who would have it. The obligation is equivalent to a moral requirement. In a disease with a low probability of penetrance and with an onset of moderate symptoms later in life, there is at most a weak, if any, reason and no obligation to prevent the life of the person who would have it. Instead, the nature of the disease would be such as to involve a permission to choose whether or not to bring about a life with it. Accordingly, there would be at most a weak reason and no obligation (permission) to terminate an embryo with a low-penetrance gene predisposing to a multifactorial form of mild late-onset hypertension. In contrast, there would be a strong reason and obligation (requirement) to terminate or not select an embryo with a high-penetrance gene causing a severe early-onset monogenic disorder like Tay-Sachs or Lesch-Nyhan.

Arguably, there would be an equally strong reason and obligation to terminate an embryo with a high-penetrance genetic mutation causing a later-onset disease with prolonged severe symptoms beginning in the middle stage of life. This would be a case where the person affected by the disease would experience its symptoms for many years before dying. Earlier-onset forms of Huntington's, Alzheimer's, and perhaps even Parkinson's might fall into this class. The reason and obligation to terminate the embryo and prevent the life in these diseases might be as strong as in the childhood diseases I mentioned because the adult with the disease might suffer at least as much as, if not more so than, the child. This would be not only a reflection of a substantial number of years of suffering, but also of the degree of suffering that comes with experiencing the decline of one's physical and cognitive capacities and a more heightened sense of dread of the future.

Others may be disturbed by the question of whether a human life is worth living. They may cite the sanctity-of-life doctrine, maintaining that all life is sacred and therefore the question of the value of a life should not even be raised. But this doctrine rejects all quality-of-life considerations and thus fails to consider the subjective experience of living with disease.[37] In doing this, it effectively bases the idea of sanctity on the biological life of human organisms rather than on the psychological life of persons. It is our psychological life that matters to us and, in some instances, pain and suffering in this life may be so severe that it would not be worth having. This suggests a further important point. When considering whether the life of a potential person would or would not be worth living, it is not any burden on the parents, but instead the burden of disease on the person who would have to live with it, which should inform our moral reasoning.

Giving Priority to the Worse Off

Two additional objections might be raised against my arguments and the principles on which they rest. Specifically, the impersonal comparative principle could be interpreted in such a way as to require us to bring into existence only people with no diseases or disabilities of any sort over the course of their lives. It even might require us to prevent the lives of people with only a genetic predisposition to certain diseases. Moreover, it could lead to discrimination against people with disabilities. These objections can be met by appealing to a particular version of egalitarianism.

Equality designates a relation between people or different groups of people. It is invoked when we want to evaluate how some people compare with others with respect to the distribution of goods. Egalitarianism is concerned with the distributive and comparative question of *who has what*, as distinct from the utilitarian concern with the aggregative question of *how much there is*. There are two interpretations of egalitarianism in the philosophical literature. One says that equality as such has value, and that inequality of any sort in the distribution of goods is morally objectionable. The other says that equality as such is not what matters morally, but instead that priority should be given to the needs and claims of the worse off.[38] On this view, inequalities are morally permissible provided that they work to the benefit of the least advantaged individuals or groups. The second version of egalitarianism is captured in John Rawls's difference principle (DP) and maximin rule (MR).[39] DP states that inequalities are to be arranged so that they are to the greatest benefit of the least advantaged. MR states that, in ranking alternatives, we should adopt the alternative the worst outcome of which is superior to the worst outcome of others. Together, DP and MR yield the worse-off priority principle. For present purposes, the worse off are people with severe disease and disability, while the better off include people with moderate to moderately severe disease, as well as those who are by all accounts healthy.

Regarding health status, "better off" and "worse off" may be construed in comparative terms involving how one group fares relative to another. Or they may be construed in absolute terms, where one is better or worse off with respect to an absolute baseline of a decent minimum level of adequate mental and physical functioning. The lower an individual falls below the baseline and thus the worse off he is in terms of disease, the stronger will be his claim to have his health needs met with appropriate medical treatment. Conversely, the higher an individual falls above the baseline and thus the better off he is in terms of health, the weaker will be his claim to medical treatment. Beyond some point above the critical level, one's claim to medical treatment will have little, if any, moral force. As I will argue in chapter 3, medical intervention for people whose health status falls above the critical level should be considered enhancement rather than treatment and pertains to people's preferences rather than to their needs. Claims based on preferences have much less moral force than claims based on needs.

Fairness is one feature of the second, "prioritarian" version of egalitarianism. A fair distribution of a good like medical treatment requires that different people's health needs be met in proportion to their strength. Fairness is a relative matter, a matter of how a person making a claim is treated relative to others making similar claims. Obviously, people with severe disease and disability have more urgent needs and therefore stronger claims to have these needs met. Consequently, a fair distribution will be one in which priority is given to the more urgent needs of the worse off (sick) over the less urgent needs and preferences of the better off (healthy).

On the prioritarian version of egalitarianism, a smaller benefit to the worse off morally outweighs a greater benefit to the better off, and a greater loss to the better off can be morally outweighed by a greater gain to the worse off.[40] Still, what matters above all is not how one group fares compared with another, but instead how people fare with respect to an absolute baseline of adequate mental and physical functioning. The point of medical intervention for the worse off is to meet their medical needs so that they can reach, or at least approximate, the critical functional level. If a treatment fails to meet this goal and only marginally benefits the worse off, and if as a consequence it indirectly causes the health of the better off to fall below the baseline, then we can give less weight to the claims of the severely diseased or disabled. This weight will be a function of the number of people who stand to benefit or be burdened in their health status in absolute terms, as well as the extent of the benefits and burdens. It is also a function of the fact that medical care is a limited resource that cannot be distributed in a way that meets all the needs of all the people making claims to receive it.

The principle that the medical needs of the worse off with severe disease and disability should be given priority is not an absolute but a prima facie one that may be overridden, depending on the distribution of burdens and benefits across all people with medical needs. Some philosophers have argued that priority to the worse off can be outweighed by substantial benefits to a substantially large number of those better off.[41] Yet what we should say here is that priority to the worse off can be outweighed if it would result in substantial burdens to the better off, when these burdens make them fall below the baseline of adequate functioning. Priority to the worse off also can be outweighed by benefits to those who are comparatively better off, but not well off in absolute terms, when these benefits raise them to the baseline. In general, preventing harm in the form of burdens to some matters more morally than benefiting others. But with respect to health, the moral weight of a claim to receive medical treatment will be a matter of degree, depending on how far above or how far below the critical level one's health status falls. Claims to receive treatment will be stronger when treatment raises people's functioning to, or maintains it at, the critical level.

Unfortunately, in most of the diseases I have cited, there are no medical treatments that can control them or alleviate their symptoms. In some cases, treatments can make people affected by disease slightly better off in absolute terms, but not raise their mental or physical condition to a level of adequate functioning. Gene therapy has shown some promise in ameliorating some conditions, as has been noted regarding severe combined immunodeficiency (SCID). But the treatment is very expensive. Enzyme replacement therapy, which is similar in some respects but not identical to gene therapy, has achieved some success in treating the early-onset Gaucher's disease, which affects 1 in 450 Ashkenazi Jews.[42] In Gaucher's, delivering the missing enzyme necessary to prevent the accumulation of fatty substances in the brain, liver, spleen, and bone marrow costs about $10,000 per month. We can project that the monthly cost of gene therapy for various disorders would be at least this much. In a health care system where allocation decisions have to be made to control inflation of medical costs, one might reasonably ask whether we would benefit more people by using the resources otherwise meant for gene therapy on less expensive and more effective programs for treatable disorders like asthma, which affect many more people, or on public health prevention programs and better prenatal care. Many more people with moderately severe disorders that make them fall below the baseline of adequate health functioning, but who are better off than the smaller number of people with severe disorders, collectively may have a stronger claim to receive medical care than the worse off. This is because benefits to them can raise their level of functioning to the critical level. Ensuring that as many people as possible reach or remain at this level is medically and morally the most important aim in the distribution of health care.

Priority to the worse off over the better off with the aim of improving health status, therefore, is not absolute but conditional on the number of people who stand to benefit or be burdened and the extent to which the people in each group will benefit or be burdened. Assuming that people with the most severe early-onset genetic diseases are not likely to benefit much from any treatment, and that spending money on expensive treatments for them means not spending it on less expensive treatments that can benefit a larger number of people, the medically and morally preferable course of action is, not to deny treatment to the worse off, but to prevent them from coming into existence. With severe and untreatable diseases, preventing the existence of the people who would have them is the only way to prevent the experience of pain and suffering.

This invites the objection to the impersonal comparative principle. I briefly addressed this objection earlier; but it warrants a more sustained discussion. Recall that the principle says that, other things being equal, it is worse to cause a person to exist if it would be possible to cause a different, better off, person to exist instead. If a prospective parent or parents had access to reproductive technology allowing them to produce multiple IVF embryos, and they could select a genetically normal embryo that would develop into a healthy person over a genetically defective one that would develop into a diseased person, then it seems to follow that they would be obligated always to select the embryo with the best genetic profile. If so, then it also seems to follow that they would be obligated to select against embryos that would develop into people with only moderate disease or disability. The suggestion is that we always should select only embryos with the *best* predictable health status, not simply those with *adequate* predicable health status. This conclusion, which is motivated by perfectionism, threatens to undermine the moral asymmetry thesis concerning genesis choices, which is motivated by nonmaleficence.

But the impersonal comparative principle does not have this implication. The moral imperative to give priority to the needs of some over the needs of others holds with respect to an absolute unit of measure. The claims of some who would be better off regarding the baseline of adequate functioning once they exist have more or less moral force depending on whether they fall below or above the baseline. By this line of reasoning, if one embryo had no genetic abnormality, and a different embryo had an abnormality entailing susceptibility to a chronic but treatable condition like mild hypertension later in life, then any medical or moral reasons for choosing one embryo over the other would be very weak. Both people developing from these embryos would be at or above the critical level of health for the balance of their lives. We would not be morally required to select the better embryo, since there would be no significant difference in the potential for harm between the two. However, we would be morally permitted to choose the better of the two. On the other hand, if the choice were between an embryo with no genetic abnormality and one with an abnormality that would result in a severe and untreatable disease, then we would be morally required to select the better embryo. This is be-cause of the likely harm that would result from allowing the genetically defective embryo to become implanted and develop into a person who would experience pain, suffering, and limited lifetime opportunities.

Another objection might come from disabilities rights advocates. They might argue that intervention in the form of testing and selectively termi-nating embryos with genetic mutations causing disabilities would reduce the number of people with disabilities. Consequently, public support for people who already have disabilities would erode. This would lead to a de-valuation of the lives of the disabled and to social discrimination against them.

To meet this objection, we can appeal to Allen Buchanan's point that "it is not the *people* with disabilities which we devalue, it is the *disabilities*."[43] One could respond by saying that people with disabilities identify with their disabilities and therefore the two cannot be so neatly separated. Nev-ertheless, a physical disability is a biological state of the body. Personhood is a psychological condition consisting in the capacity for mental states such as consciousness, memories, intentions, and beliefs. While the biolog-ical condition of our bodies and brains largely influences the nature of our mental states, these states are neither identical nor reducible to biological states. Persons are constituted by their bodies and brains, but are not iden-tical to them. An argument by Lynn Gillam supports this position:

> [T]hese unfavorable quality-of-life assessments cannot, strictly speaking, be regarded as discriminatory towards people with disabilities. This is because they are not applied to people now living, but to fetuses, to ground a decision about what may be done to a fetus. This is the crucial point, because on the understanding of abortion considered here, fetuses are not persons, and moral decisions about fetuses cannot logically be ex-tended to persons. [44]

Persons have disabilities; embryos and fetuses do not. Since embryos and fetuses are not identical to persons, and only persons can be harmed, ter-minating an embryo or early-stage fetus that would develop into a person does not harm anyone. So Buchanan's point that disabilities and the people who have them are conceptually distinct can be sustained.

Buchanan says further:

> We devalue disabilities because we value the opportunities and welfare of the people who have them—and it is because we value people, all people, that we

care about limitations on their welfare and opportunities. We also know that disabilities, as such, diminish opportunities and welfare, even when they are *not so severe* that the lives of those who have them are not worth living.[45]

The underlying rationale for this view is that removing or preventing limitations on a person's opportunities for a decent life is a matter of justice, not only of nonmaleficence or beneficence. Yet the second passage cited from Buchanan leaves open the possibility that we would be morally required to terminate embryos known to have genetic defects that would lead to people having lives with limited opportunities, but which nonetheless would be worth living. Against Buchanan, the arguments presented in this section support the claim that we should terminate only those embryos with genetic defects manifesting in severe diseases and disabilities that make lives on balance not worth living.

Disabilities rights groups have been successful in securing legal protection through political means, as the Americans with Disabilities Act of 1990 firmly attests. But to maintain that preventing additional people who would have disabilities from existing would lead to discrimination against people who already have them suggests that those coming into existence with disabilities might be treated instrumentally as mere means for the benefit of others. While this may yield political dividends for those who already exist, it would be difficult to justify on moral grounds. Moreover, one could argue that the smaller the number of people with severe disabilities, the better able society will be to meet their needs. For example, it might help to ameliorate the problem that many school systems have in paying for nurses, physical therapists, psychologists, and special buses for severely disabled children, without sacrificing the less urgent but still significant educational needs of many other students.[46]

More controversially, if the burdens to a substantial number of people morally outweigh the benefits to one person or a few people, then there may be a moral obligation to prevent a life that would involve only a moderate or moderately severe disease such as CF. This could enter into parents' deliberation about whether to bring a diseased child into existence when they are obligated to meet the less urgent needs of children they already have. In a family unit, the needs of all the members matter morally. Although priority should be given to meeting the needs of a family member who is worse off than her siblings or parents because of poor health, in some instances this priority can be overridden. This would be the case if priority to the worst-off member did little to improve her condition and resulted in substantial burdens to all other family members. The distribution of needs among all family members may make it impossible to adequately meet the needs of the worst-off member. Rather than cause a diseased child to suffer in such a scenario, it might be preferable to prevent any harm to her, and to other family members, by preventing her from existing.

Thomas Nagel once offered the following example to support the idea of giving priority to the worst-off member of a family.[47] The parents of two children, one healthy and happy, the other suffering from a painful handicap, decide to move to a large city from the suburbs. In the city, the handicapped child will be able to receive the medical treatment and special schooling that he urgently needs, neither of which would be available to him if he and his family were to remain in the suburbs. As a consequence, the healthy child will not be as happy as he would have been remaining in the suburbs, and the family's overall standard of living and quality of life will be lower in the city. The strength of the egalitarian claim for special treatment for the handicapped child depends on his worst-off condition relative to his sibling and parents and with respect to the absolute baseline of adequate functioning he needs to reach. He will be able to function physically and mentally better in the city. His needs are urgent enough that they outweigh the less urgent needs of his sibling and parents.

But suppose that the handicapped child has six siblings, and that each of these six, together with each of the parents, would be burdened by living in the city. It seems unfair to insist on giving absolute priority to the needs of this worst-off child when the resulting collective burden on the rest of the family will be substantial. If there are many family members, then, even if their individual burdens fall short of the burdens the handicapped child must endure, the collective burden on the family unit as a whole could morally outweigh the burden on the child. This could weaken the strength of his claim to have his needs met.

Imagine that prospective biological parents are both carriers of a recessive trait that causes a moderate to moderately severe disease. They have adopted six children out of fear that any child they conceived through natural biological means would have a 25% chance of inheriting the trait and developing the disease. If they know that an IVF embryo they produced had the mutation that would cause the disease, then they might be obligated to terminate the embryo and not have the child who would have developed from it. This is because of the burden that meeting the needs of this child would have on their other children and themselves, despite the fact that the disease would be only moderate to moderately severe. The parents' obligation to the welfare of their existing children would ground their obligation to prevent the existence of the additional child. Although the expensive procedure of producing IVF embryos would not be a likely scenario here, parents would have stronger moral grounds on which to make such a decision the poorer they were. Their economic situation would increase the overall burden on the family.

Conclusion

I have focused on the epistemic aspect of moral obligation and responsibility and have examined our obligation to the people we bring into existence. First, I addressed late-onset diseases of familial inheritance such as Huntington's and breast cancer attributed to mutations in the BRCA1 and BRCA2 genes. I noted the differences in the probability of genotypic penetrance in these two diseases. Then I considered whether persons believing themselves to be at high risk for these diseases, or knowing they have them given certain symptoms, have an obligation to be tested to determine whether they have the mutations that cause the diseases and to share that information with their children or siblings. The moral reason for genetic testing and sharing genetic information is to prevent harm to others who also are at high risk. The obligation to do this is stronger the higher the probability of penetrance, the more severe the disease, and the greater the number of people who may be affected by it. There may be prudential reasons for being tested as well. Having genetic information could enable one to plan one's life so as not to lose any opportunities and may enable one to take steps to prevent avoidable pain, suffering, and even premature death.

I then examined early-onset monogenic diseases and considered whether presymptomatic or carrier parents who know the risks are obligated to test embryos or early-stage fetuses for mutations that cause these diseases and terminate their further development. Ideally, if these parents have affordable access to genetic reproductive technology, then they could produce multiple IVF embryos, have them tested for a genetic abnormality, and on the basis of a preimplantation genetic diagnosis select an embryo without the abnormality for implantation and development into a disease-free person. Parents are morally obligated to terminate or not select embryos whose cells contain genetic defects that cause prolonged severe disease or disability in people who develop from them. I explained obligation as a matter of degree, falling along a continuum between moral requirements and moral permissions. The obligation to prevent a life will be stronger or weaker depending on the following four factors: the probability of genotypic penetrance; the severity of the symptoms associated with the disease; the time of onset of symptoms; and the time between onset of symptoms and death.

The moral justification for this view is that it is wrong to cause people to exist when the pain and suffering they experience make their lives not worth living on the whole. This claim is motivated by the principle of nonmaleficence, as well as by the person-affecting and impersonal comparative principles. If we cause people to exist with severe disease and disability, then we defeat their interest in living without pain and suffering and thus harm them. More generally, when we are considering whether to bring people into existence, it is better, other things being equal, to prevent the existence of a person who would have a severe mental or physical disease and instead bring into existence a person with normal mental and physical functioning. By doing so, we avoid adding to the total amount of suffering in the world. In addition, the principle of justice supports this position. People will not come into existence with a condition that would severely limit their opportunities for achieving a decent minimum level of lifetime well-being. All of these claims are sustained by the fact that, for most severe genetic diseases, there are no treatments that can control them or alleviate their symptoms.

I also argued that sometimes we may be morally obligated to prevent the existence of people who would have only moderate to moderately severe diseases and would have lives worth living. This could occur in families with many members, where the burdens to siblings and parents could collectively outweigh the benefits the affected child would receive in the form of expensive medical and other special treatment. Such a scenario might arise in the case of a potential child who would have CF. We should give priority to the more urgent needs of the worse off over the less urgent needs of the better off. But this priority is not absolute. Substantial burdens to a substantial number of people may outweigh the claims of the worse off to be benefited. Rather than harm them by not providing them with the treatment they need, it may be morally preferable not to bring into existence people who would have severe untreatable diseases.

Many of the issues that I have discussed in this chapter fall into the category of what Lee Silver calls "reprogenetics,"[48] or the merging of reproductive biology and genetics. But genetic testing of embryos, preimplantation genetic diagnosis, and embryo selection are only some of the genetic interventions affecting the number, identity, and quality of life of the people we bring into existence. Gene therapy and genetic enhancement of physical and cognitive traits are other important interventions. In the light of the genethical issues that I just mentioned, we need to examine and evaluate these and other genetic interventions in some detail.

3

GENE THERAPY AND GENETIC ENHANCEMENT

The idea of genetic intervention often suggests inserting into cells a normal copy of a gene coding for a protein crucial to the normal function of these cells and other bodily processes. Ideally, diseases with a genetic cause could be controlled or even cured. In this regard, genetic intervention falls within the general goal of medicine, which is to prevent or treat disease and maintain or restore people's physical and mental functioning at or to normal levels.[1] Testing and selectively terminating embryos with deleterious genetic mutations that otherwise would develop into people with severe disease and disability is a form of prevention, while gene therapy for people who exist with genetically caused diseases is a form of treatment. Each of these two types of intervention is grounded in the moral principles of beneficence, nonmaleficence, and justice. They aim at avoiding harm to people and benefiting them by allowing them to realize their interest in having healthy lives. And they aim at ensuring fair equality of opportunity to achieve a decent minimum level of lifetime well-being.

Genetic enhancement is a third type of genetic intervention. Its aim is to improve one's phenotypic traits and thereby raise one's physical or mental functioning above normal adequate levels for persons. This is significantly different from genetic testing and termination of embryos and gene therapy, not only because its aim falls outside the goal of medicine and health care, but also because it is motivated by perfectionism rather than by beneficence and justice. In fact, many would say that, as a program of positive eugenics, genetic enhancement is antithetical to beneficence and justice precisely because of the pervasive social inequalities that would result from its practice on a broad scale.

Yet, others assert that it is naive to think that treatment of disease to maintain or restore proper physical and mental functioning is categorically distinct from enhancement of this functioning. An intervention considered as an enhancement in one respect could be considered as a treatment in another. If this is correct, then there is no clear separation between treatment and enhancement, but only a continuum along which they differ only in degree and not in kind. There would be no ground on which to say that only the first and not also the second type of intervention is morally justified. The moral distinction between beneficence and justice, on the one hand, and perfectionism, on the other, collapses. It seems that moral justifications for one or the other type of genetic intervention stand or fall together.

But I will argue in this chapter that we have a sound general understanding of what constitutes adequate or normal physical and mental functioning. On this basis, we can clearly distinguish between interventions that restore people to or maintain them at a normal functional level, and interventions that raise them above this level. Furthermore, I will argue that medical and moral reasons for genetic intervention have more weight at and below the baseline of adequate functioning and less weight above the baseline. This is because the weight of the reasons supporting claims to receive treatment is proportional to the needs of people making these claims, and enhancements do not pertain to people's needs but to their preferences above the critical level where their basic health needs have been met.

First, I discuss gene therapy and explain why it has yet to become effective for treating most genetic diseases. This will serve to reinforce my general argument in chapter 2 regarding the prevention of the lives of people who would experience severe pain and suffering because of these diseases. I distinguish somatic-cell gene therapy from germ-line genetic alteration. Elaborating further on the discussion of allocating scarce medical resources, I consider whether we should give higher priority to treating people with severe but rare monogenic diseases, or to people with less severe but more common multifactorial diseases. The number of people who stand to benefit from treatment, as well as the extent to which they benefit, affect how equality and fairness are evaluated in this context. I also examine the differences between gene therapy for physical disorders and gene therapy for cognitive disorders and explain why the latter would have a significant impact on personal identity and our understanding of beneficence. Shifting to genetic enhancement, I argue that there are neither medical nor moral reasons for this type of intervention. I give five reasons why genetic enhancement should be prohibited. Finally, I address the claim that there is a slippery slope from treatment to enhancement, or that the negative eugenics corresponding to prevention and treatment inevitably slides into the positive eugenics corresponding to enhancement. I maintain that this claim can be rejected because the slippery-slope argument on which it rests is unsound, owing mainly to the fallacy of assimilation and unsupported claims about causation in its premises that render them false. Provided that the aim of genetic intervention is prevention or treatment of disease, and provided that adequate policy guidelines and legislation are in place, negative eugenics can be medically and morally justified.

Gene Therapy: Problems and a Paradox

Roughly, gene therapy consists in the correction of a mutated gene or the insertion of an additional normal gene into a person's somatic cells to treat a disease with a genetic cause. In contrast, genetic alteration of germ cells (gametes: sperm and egg) at the embryonic stage of a human organism is a form of disease prevention. This second type of genetic intervention should not be considered "therapy," because when the gametes are altered there is no existing person who might benefit from this procedure.[2] "Therapy" implies that there is a disease to be treated, and it is not embryos or early-stage fetuses but persons (and, arguably, late-term fetuses) who have diseases and their respective symptoms. Defective alleles of genes in embryonic cells may be the causes of diseases, but diseases and their symptoms do not manifest themselves until the organism has reached a further stage of development. Like germ-line genetic alteration in embryos, testing preimplantation embryos for genetic abnormalities and selecting those with abnormalities for termination is a preventive rather than a therapeutic strategy. It is designed to prevent severe diseases by preventing the existence of the people who would have them.

Somatic-cell gene therapy may determine or preserve the identity of a person, depending on the stage of development of the organism when the intervention takes place. It also depends on whether the trait or function treated is physical or mental. In chapter 1, I said that the addition of a normal copy of the gene to treat Ashanti de Silva's ADA-SCID would not likely alter her identity. For it would not significantly alter her memories, beliefs, desires, and intentions or the ways in which they were interconnected. However, somatic-cell gene therapy designed to correct or treat a cognitive or affective disorder would be more likely to alter one's identity.

The manipulation of the relevant neurotransmitters or regions of the brain that generate and support mental life would directly affect the very nature of the mental states definitive of personhood and personal identity through time. The gene therapy for the two French babies with a different form of SCID, also mentioned in chapter 1, would be identity-determining as well. Although this disorder is physical rather than mental, the intervention took place before the babies had developed a unified set of mental states to be full-fledged persons. But whereas the physical therapy in this case would be identity-determining because it occurred at such an early stage of development, cognitive therapy for a long-term mental disability most likely would be identity-determining at any stage of one's psychological life. As I will explain, this should lead us to question whether interventions to treat cognitive disorders could be considered therapeutic.

Germ-line manipulation would determine rather than preserve personal identity because it would entail a completely different set of biological and mental properties of the person who comes into existence after the manipulation. This individual would be distinct from the one who would have existed had the manipulation not taken place. The biological properties would be completely different because the intervention would occur before the process of cell differentiation had begun, or while it still was in progress. Even slight changes to the biological properties of a human organism at the embryonic stage would cause different psychological properties to develop and thus select between which of a number of different people would come into existence.[3] The same could be said about in utero fetal gene therapy. Although the insertion of a normal copy of a gene to treat a genetic abnormality would occur after most cell differentiation had taken place and incipient organ systems had formed, the biological and physical life of the person who came into existence would be different from the one who would have existed without the therapy. And whether one exists with or without a physical disease or disability will determine to a significant extent the nature and content of the mental states in virtue of which an individual is and remains one and the same person. Somatic-cell and germ-line genetic manipulation at an early stage of development would have the same effect on personal identity. Still, these two types of intervention differ in at least one crucial respect. The first determines the identity of one person as distinct from another person who would have existed without it. The second can determine the identities of many future people, given that the effects of the manipulation are passed on to offspring.

Somatic-cell gene therapy is not yet feasible for treating the majority of genetically caused diseases. For one thing, gene therapy is primarily relevant to single-gene disorders, and polygenic diseases such as colon cancer involve several genetic mutations. Furthermore, chromosomal disorders such as Down syndrome involve large segments of DNA that are not amenable to treatment. In addition, the monogenic diseases that might be amenable to gene therapy are recessive rather than dominant.[4] Recessive disorders—where a single copy of a normal allele is sufficient for functioning within the normal range for persons—could be treatable because inserting the copy into cells containing a double dose of the abnormal allele could be enough to ensure proper cell function. Dominant disorders, on the other hand, result even when an affected individual has only a single copy of the defective allele in his or her cells. These disorders would be treatable only by correcting the mutation in that allele. Gene replacement would be superior to gene addition because its potential to correct a mutation could prevent both autosomal dominant and recessive diseases.[5] Presently, however, it is not possible to correct a mutation but only to insert an additional copy of a gene into the affected cells. Only a few autosomal recessive and X-linked disorders can be treated with gene therapy.

In theory, a recessive disorder like cystic fibrosis would be treated by delivering a normal copy of the critical allele into the affected cells so that the pancreas produced the necessary enzymes for food absorption and the glands in the lining of the bronchial tubes functioned properly. In an X-linked disorder like hemophilia, a normal copy of the gene coding for the protein that controls blood clotting could be delivered through injections. In practice, what has plagued the efforts of gene therapy to control or cure disease has been the lack of a suitable vector to deliver therapeutic genes into cells and maintain them in proper working order. Viral vectors have been the most common method used to date, specifically inactivated adenoviruses and retroviruses used to carry pieces of DNA into cell nuclei. Yet this method has been largely unsuccessful because these stripped-down viruses do not provide a stable platform for the genes to operate efficiently. The fact that the respiratory tract consists of rapidly dividing epithelial cells means that frequent injections of the vector carrying the gene would be needed. Some of these viruses are not large enough to carry a full human gene and all of its "switches," while other viruses may provoke an adverse response from the immune system. In addition, if the vector were not targeted precisely to the gene or genes in question, then that vector could adversely affect other genes and interfere with their functions, which could lead to various cancers. Or the delivery virus might revert to its normal active state if it interacted in a certain way with natural viruses in people's lungs. Finally, a person could develop immunity to the viral vector, in which case the treatment would cease to be effective.[6]

The risk of harm to individuals undergoing gene therapy has become palpably evident in the wake of three recently reported deaths resulting from complications of viral vectors. Jesse Gelsinger had a mild form of ornithine transcarbamylase (OTC) deficiency, whereby the liver cannot process ammonia, a toxic breakdown product of food. He died in 1999 following gene therapy with a viral vector intended to correct the disorder at the University of Pennsylvania.[7] Several months later, it was reported that James Dent, who had been diagnosed with a terminal form of brain cancer, underwent gene therapy in Toronto and died from complications of the treatment.[8] Also, an individual died after receiving gene therapy involving vascular endothelial growth factor (VEGF) at St. Elizabeth's Hospital in Boston.[9] Citing these deaths, many physicians, researchers, and ethicists have insisted that there be a moratorium on gene therapy until the procedure has been perfected and the risks minimized. In the meantime, they argue, gene therapy should be reserved only for terminal patients as a last resort.

Alternatively, liposomes might be safer and more effective vectors to deliver the crucial genes than viruses, though this has yet to be borne out by the research.[10] Pharmacogenomics would be an even more promising method, since the therapeutic drugs used would target the relevant genes more specifically and would ensure a more balanced interaction among genes, drug toxicity, and metabolism.[11] But this is far from being perfected as well. As it stands, gene therapy has shown some success in treating some people with hemophilia, SCID (as in the cases of Ashanti de Silva and the French babies), and familial hypercholesterolemia. Particularly significant is a recent study showing that nonviral gene therapy for hemophilia A, using plasmids as vectors, has been successful in producing factor VIII, which is essential to blood clotting.[12] Also, research is well under way to develop in utero gene therapy on second-trimester fetuses for alpha-thalassemia.[13] For the majority of monogenic and multifactorial diseases, however, gene therapy does not yet offer an effective treatment or cure. While it may offer these in the future, for now genetic testing and selective termination of embryos with mutations that cause severe genetic diseases seems to be the most effective way to prevent the severe pain, suffering, and restricted lives these diseases entail.

Somatic-cell or germ-line genetic manipulation of embryos may appear more attractive than testing and terminating them, as well as more effective than gene therapy given at later stages of human development. Intervention at such an early stage could prevent the disorder without having to prevent people from existing. If the manipulation involved deleting a mutation at the germ line, then it would eliminate the mutation from the human population, since it would not be passed on to offspring. Among other things, however, germ-line genetic alteration may not be desirable from an evolutionary perspective. Some genetic mutations are necessary for species to adapt to changing environmental conditions, and some genetic disorders involve alleles that confer a survival advantage on certain populations. As noted earlier, the best example of this phenomenon is the allele causing sickle-cell anemia, which provides greater resistance to malaria for populations from a region in Africa. Altering a gene or genes at the germ line to correct one disorder may only lead to other disorders that could adversely affect people in many successive generations.

This risk raises the question of whether we have a duty to prevent passing on altered genes with potentially harmful consequences to people who will exist in the distant future. It may recommend avoiding germ-line genetic manipulation altogether, which is supported by two related points. First, people existing in the future may be adversely affected by the consequences of a practice to which they did not consent. Second, because of the complex ways in which genes interact, it would be difficult to weigh the probable health benefits of people in the present and near future generations against the probable health burdens to people in the distant future. Because their interest in, and right to, not being harmed have just as much moral weight as those of the people who already exist or will exist in the near future, we would be well-advised to err on the side of caution. Indeed, we would be morally obligated to do so, on grounds of nonmaleficence. This would mean prohibiting germ-line genetic manipulation, or at least postponing it until further research can provide a more favorable assessment of its safety and efficacy.[14]

There is one form of germ-line intervention, however, which might not occasion the same concern as others. Ooplasmic transplantation consists in introducing mitochondria from younger women into the eggs of older women to increase the success rate of IVF infertility treatment. This is germ-line intervention because the DNA from the two mothers is passed down the maternal line to future generations. Because it is confined to the mitochondria, presumably it would not change the DNA in the nucleus of an egg. But it is still not known whether changes to mitochondrial DNA in ooplasmic transplantation could effect changes in nuclear DNA. Because it is difficult to assess the risk–benefit ratio *ex ante* in interventions with effects that are passed on to offspring, there may be biological reasons against this form of germ-line intervention as well. Moreover, it is questionable whether such an intervention could be called gene therapy, since it is questionable whether the condition it is used to treat could appropriately be considered a disease.

With respect to somatic-cell gene therapy, it is instructive to consider further how it bears on the metaphysical question of personal identity and on moral questions about benefit, harm, equality, and fairness. Recall the example in chapter 2 about the boy with DMD who files a tort of wrongful life against his parents, obstetrician, and/or genetic counselor on the ground that they were negligent in assessing the risk of a child having the disease. There would *not* be grounds for a claim or tort of wrongful life, however, if the child insisted that genetic intervention other than testing and selective termination of the embryo from which he developed should have occurred. For example, he could not claim that delivering a normal copy of the dystrophin gene at the embryonic stage would have made a significant difference between the life he actually has and a better life he might have had. For at the time of this intervention, there was no identifiable person who could have benefited from it. Genetic intervention at such an early stage of development of a human organism would have resulted in a completely different set of biological and psychological properties belonging to a different person. So the child only could claim that the harm and wrong were committed by not terminating the genetically defective embryo from which he developed, not by failing to add a normal copy of the dystrophin gene.

Gene therapy for a physical disorder involving somatic cells at some age considerably after birth probably would mean some difference in biological and psychological properties before and after the intervention. But the difference likely would not be so great as to disrupt psychological connectedness and continuity and to determine the identity of a distinct person. We can assume that the individual already would have lived for a certain number of years and would have a fairly well-developed biological and psychological life. Although some mental and bodily properties would change as a result of genetic intervention, there would be enough similarity between these properties before and after the therapy to say that a child cured of CF or SCID, for instance, would remain the same person. If the parents of a child with either of these diseases had affordable access to genetic technology that could cure the disease or control its symptoms, but failed to avail themselves of it, then perhaps the child could claim that her parents harmed her by defeating her interest in not experiencing a disease that could be cured or controlled through gene therapy. But to the extent that the child already exists with a life worth living, and that her parents do not have control over and thus are not responsible for the disease, this case would involve something other than the standard notion of wrongful life.

On the controversial assumption that second-trimester fetuses are persons, fetal gene therapy could be either an identity-determining or identity-preserving form of genetic intervention. W. French Anderson maintains that "fetal gene therapy offers a powerful technology that could correct genetic diseases which now produce irreversible damage before birth, so avoiding the need for alternative techniques such as gamete donation, embryo selection, abortion, adoption, or non-parenthood."[15] The intervention would take place after much cell differentiation already had occurred. But if the disease that is cured were severe to moderately severe, then correcting the gene that caused it would result in a significantly different set of biological and psychological properties after birth. The psychological properties would be especially crucial, since their nature and content could be quite different depending on whether one experienced a healthy life or a severely diseased one. So whether the same or a different person existed after the intervention would hinge largely on the severity of the disease in question. But all of this rests on the assumption that second-trimester fetuses are persons, an assumption that I do not make.

All of the cases that I have discussed thus far involve physical diseases. More difficult to assess would be gene therapy to correct or treat a mental disorder at any stage of a person's life. Recently, researchers reported that they were able to manipulate a gene in the brains of mice and thereby enhance their memory.[16] This suggests the possibility of restoring or even improving cognitive function through manipulation of the analogous gene or genes in humans. Quite apart from the obvious differences between mice and humans, it is simplistic to think that a single gene or gene product could control intelligence. This mental capacity is a function of complex interactions among different genes, whose functions in turn are affected by the interaction of human organisms with their social and physical environment. Still, subtle alterations in brain biochemistry can affect the nature of and connection between mental states such as beliefs, intentions, and emotions in which personhood consists. Manipulating a gene that plays a crucial role in cognition or emotion could affect these mental states and therefore alter personal identity. Although genetic intervention to treat mental impairment is not likely in the foreseeable future, it is instructive to test our metaphysical and moral intuitions about persons by considering the respects in which genetic intervention could alter one's cognitive and emotional states.

Whether these interventions preserved or changed the identity of persons would depend on the purpose of the intervention and the nature of the mental disorder in question. If a person were to lose previously normal mental functioning as a result of a stroke, then the purpose of any genetic treatment (for example, angiogenesis to restore blood flow and neuronal connections) on the relevant regions of his brain would be to restore the level of mental functioning he had before the stroke. While there would be a gap in psychological connectedness and continuity during the period when he was experiencing the effects of the stroke, the intervention ideally would restore connectedness and continuity and hence his former self. On the other hand, cognitive or affective treatment for someone with moderately severe to severe mental impairment from birth due to a genetic or chromosomal anomaly, such as Fragile X or Down syndrome, would be aimed at raising their level of mental functioning to what is considered normal for persons.

In each case, the genetic intervention would have a therapeutic rationale. But it would have a different impact on personal identity. In the first case, the restorative effect of the intervention would reestablish the connections between the normal mental states the person had before the stroke with the normal states he would have following the treatment. There would be some disconnectedness between the time of the stroke and the time the treatment restored his mental functioning. Yet this would be temporary rather than permanent. Provided that the temporal gap was not too long, the intervention would restore his mental functioning and identity and would appropriately be called "therapy." It is "therapeutic" because it makes one and the same person better off than before the stroke. In the second case, the nature and content of the individual's mental states would be radically different before and after the genetic intervention and thus would mean that the original person with the mental disorder became a distinct person. Even if the memories of the individual's past experience remained intact after the therapy, how he reconstructed and integrated them into a unified set of mental states would be influenced by more robust intentions and desires. These in turn would make his present state of consciousness qualitatively different from what it would have been without the intervention. If this account is correct, then the intervention could not be therapeutic. For the person who presumably benefited from it no longer would exist. The treatment for the mental disorder would alter the mental states of the person to such a degree that the identity of the person with the disorder would be altered as well.

The intervention in the stroke case could be defended on grounds of beneficence and justice. By restoring the person's mental states to the normal level of functioning he had before the stroke, we would be benefiting him by making him better off than he otherwise would have been with its adverse mental effects. Moreover, correcting the dysfunction would restore his level of functioning where he had the same general opportunities as other people to undertake and complete projects within a life plan of his own making.[17]

In contrast, if we could deliver a gene or genes into the brain of a severely cognitively or affectively disabled individual and dramatically raise her mental functioning to a normal level, then we would not benefit *this* person. She would not benefit from the intervention and be made better off because the change in her mental states would mean a change in identity. The intervention would not be therapeutic because it would not be the case that one and the same person was made better off. At first blush, the justice requirement would offer stronger reasons for improving the mentally disabled person's cognitive or affective capacities through genetic means. A person should have opportunities for projects and achievements of which she is deprived because of her mental disability, opportunities equal to those of people with normal mental functioning. Yet the difference in identity before and after the intervention suggests that the justice requirement would pertain more to the level of cognitive or affective capacity than to the particular persons who have or lack these capacities to varying degrees. It would not matter morally, one might argue, that a particular mentally disabled person became a different person following genetic intervention, only that some person had the same level of mental functioning and the same opportunities for achievement as other persons.

This suggests a utilitarian argument for justice. That is, raising the aggregate level of mental functioning across all people matters morally more than the level of mental functioning of particular people. Yet, when we specify the beneficence and justice requirements, we assume that the cognitive or affective gene therapy that makes a person better off and increases a person's opportunities preserves the identity of that same person. We aim to make *her* better off than *she* was before the intervention and thus provide her with the same opportunities as other people. Only by preserving her identity could the intervention be therapeutic. To insist on the utilitarian argument that it does not matter *who* has these opportunities, only *that* there are opportunities, fails to respect the distinctiveness of persons and undermines the concepts of "benefit" and "therapy."[18] A similar problem besets those who champion the use of implantable brain chips to compensate for mental disability.[19] The argument for this intervention is based on the principle of equal opportunity. Yet altering one's mental states to the point of making one a different person conflicts with egalitarianism and undermines the aim of compensation. Beneficence and justice presuppose the idea of distinctiveness, which in turn presupposes the identity of one and the same individual persisting through time. Although the impersonal comparative principle provides some support for the utilitarian argument for genetically improving mental functioning, a more persuasive argument would have to be advanced for ignoring the distinctiveness of persons embodied in the egalitarian conception of beneficence and justice.

If the egalitarian conception of benefit and opportunity is more plausible than the utilitarian conception, then there may be reasons for *not* using cognitive gene therapy on severely mentally impaired individuals. These reasons would be stronger in the case of a moderate to moderately severe cognitive disability, as in Down syndrome. Why try to raise the level of cognitive and affective functioning and change the identity of persons if their mental capacities are sufficient for happy and fulfilling lives? To be sure, we could increase lifetime opportunities; but these could come at the expense of the identity of the person who presumably would have them. The upshot is that genetic intervention to improve one's mental functioning is therapeutic when it *restores* one's functioning to a previous normal level. But it is not therapeutic when it significantly *raises* the mental functioning of lifelong mentally impaired individuals to a normal level. Therapy presupposes preservation of personal identity. In the first case, the genetic intervention would preserve identity; in the second case, it would alter identity.

This point poses a challenge to the egalitarian idea of giving priority to the worse off, which further underscores some of the counterintuitive implications of cognitive gene therapy. Let us assume that an individual who has been cognitively impaired from birth is worse off in his life as a whole than one who suddenly has fallen below the baseline of adequate cognitive functioning due to a brain injury. If genetic intervention into the latter's brain could restore him to the baseline, and if the same sort of intervention did little to improve the condition of the former, then priority to the worse-off individual could be overridden. This is because the benefit in being restored to the baseline of adequate mental functioning is more morally significant than the marginal benefit of ameliorating the condition of the more severely affected individual who cannot be brought up to the baseline. The benefit in question pertains not to a specific time of persons' lives, the time at which the intervention might take place, but to their lives on the whole.

Recall that the worse-off priority principle says that a smaller benefit to the worse off has more moral weight than a larger benefit to the better off. But if the aim of any form of therapy is to raise or restore people to a baseline of adequate physical and mental functioning, then the significance of reaching that level for the better off may be enough for their claim to treatment to outweigh the claim of the worse off for the same. It is morally preferable to benefit the comparatively better off and restore or raise them to the baseline than to benefit the comparatively worse off if doing so only marginally improves their condition and fails to raise them to that critical level. Suppose that the condition of the worse off could be raised to the baseline of adequate mental functioning. Intuitively, this would make a strong case for giving priority to them. But if the improvement involved radical changes in the nature and content of their mental states, then they in fact would not be made better off, since they would not be the same persons after the intervention. This is quite unlike the case of restoring the better-off stroke patient to the baseline, since his mental states effectively would be reconnected as a result of the intervention and his identity would be preserved. The coherence of and justification for cognitive gene therapy hinge crucially on the preservation of personal identity and whether the purpose of the genetic intervention is to restore or raise people to a baseline of adequate mental functioning.

Access and Allocation Issues

Suppose that gene therapy were to become feasible for treating most physical and mental disorders with a genetic cause. The fact that these treatments would be very expensive would generate problems of equality and fairness. The main concern would be with equal affordable access to the technology. The issue of access also would affect equal opportunity for a decent minimum level of well-being, since having these opportunities presupposes adequate physical and mental functioning that the technology would guarantee. There would be unequal access to this technology because not everyone would be able to afford it. This would be unfair because it would give an advantage to some people over others who have an equal claim to treatment to meet an equal physical or mental need they have through no fault of their own.

One way to resolve this problem of fairness is to implement what Maxwell Mehlman and Jeffrey Botkin call a "genetic lottery."[20] Owing to its high cost, therapeutic genetic technology would be considered a scarce medical resource to be allocated by lot within limits specified by the health care system. People who needed the technology for treatment of a medical condition would participate in the lottery on a voluntary basis and would be selected randomly for access to the technology, regardless of the ability to pay or social status. Mehlman and Botkin spell out the advantages of such a lottery:

> It would accommodate the assumption that not all genetic services could be made available to everyone. It would permit continued research toward conquering genetic diseases. It would enable people, who could not otherwise afford them, access to genetic technologies. Within the lottery, at least, there would be true equality of opportunity.
>
> By permitting winners to obtain access to whatever genetic services they wished, so long as the services were available to the open market, the lottery would avoid the onerous task of deciding the technologies to which people should be given access.[21]

They go on to propose that such a lottery could be financed by a progressive income tax, which would transfer benefits from the economically advantaged to the economically disadvantaged. Alternatively, health benefits of the better off (healthy) could be taxed to subsidize access to genetic technologies for the worse off (diseased). Although Mehlman and Botkin do not differentiate gene therapy from genetic enhancement within the framework of the lottery, it should be said that only a lottery that gave people access to treatment but not enhancement would be morally defensible and fair. This is because scarce resources should be allocated to meet people's need for treatment, not their preferences for enhancement, given that needs have more moral weight than preferences. Fairness says that people's needs should be met in proportion to their strength, which is a function of degree of need. And people who require gene therapy to raise them to the baseline of adequate functioning have more urgent needs than those who prefer genetic enhancement to raise them above that level. On grounds of fairness, then, a genetic lottery would be defensible only for people whose diseases or disabilities gave them a genuine need for treatment, the need to be restored or raised to the baseline.

Another aspect of fairness is the number of people who would stand to benefit from expensive genetic technology. This should be considered in contrast to the number of people who stand to benefit from other, less expensive, treatments. As in the case of the lottery, what motivates the question is the need for allocation decisions within specified limits to control health care costs. Here, however, the genetic lottery would not be a viable model, since we would have to compare treatments involving genetic and nongenetic interventions. Health care currently constitutes a significant proportion of U.S gross domestic product (around 14 %), and in order to control inflation of this sector the costs and benefits of genetic treatments have to be weighed against the costs and benefits of other, more standard, treatments.

Consider the number of people afflicted with monogenic disorders such as Canavan disease (1 in 5,000), cystic fibrosis (1 in 2,500), Duchenne muscular dystrophy (1 in 3,500 males), and Tay-Sachs (1 in 3,600).[22] Because of the interaction between the vector carrying the gene and the targeted cells, the effect of an injected gene would be only temporary. Repeated injections would be needed. Based on the monthly cost of enzyme-replacement therapy for Gaucher's disease, we can project that gene therapy could cost as much as $10,000 per month. Compare these figures with the significantly larger number of people with conditions like hypertension, Type 2 diabetes, and asthma, which can be controlled more effectively or prevented with less expensive treatments. Some of these treatments may consist in comprehensive public health and prenatal care programs. If we were to strictly follow the idea of giving absolute priority to the worse off, then we should allocate a larger percentage of medical resources to special treatments for comparatively rare genetic diseases, owing to their severity. However, as argued earlier, priority to the worse off can be overridden if a substantially larger number of comparatively better-off people could benefit from less expensive treatments, especially if gene therapy does not significantly improve the condition of the smaller number with genetic diseases. Thus, with respect to the health of the general population, priority in the allocation of scarce medical resources should be given to standard forms of treatment and prevention over genetic treatments.

One might argue that resource priority should be given to genetic research on possible genetic treatments for chronic multifactorial conditions like heart disease or cancer, instead of monogenic conditions. Again, the rationale would be that a substantially larger number of people would benefit from this order of priority, and at roughly the same cost of treating the rare conditions. In cancers, for example, most tumors involve a mutated p53, or tumor suppressor, gene. Replacing the gene with a normal copy might suppress the growth of tumor cells. Gene therapy could control the rate of cell division and thereby prevent the growth of tumors responsible for all cancers.[23] In addition, genetic research might lead to the development of vaccines to prevent viral infections such as HIV-AIDS. With respect to genetic research, priority should be given to research and treatment that could benefit more people affected by more pervasive but less severe diseases over research and treatment that could benefit fewer people affected by rarer but more severe diseases. Prima facie, priority should be given to meeting the most urgent needs of people with the most severe diseases. But this priority can be overridden if the same amount of resources could be used to treat and significantly benefit a substantially larger number of people in terms of their health.

Genetic Enhancement

Gene therapy must be distinguished from genetic enhancement. The first is an intervention aimed at treating disease and restoring physical and mental functions and capacities to an adequate baseline. The second is an intervention aimed at improving functions and capacities that already are adequate. Genetic enhancement augments functions and capacities "that without intervention would be considered entirely normal."[24] Its goal is to "amplify 'normal' genes in order to make them better."[25] In chapter 1, I cited Norman Daniels's definitions of health and disease as well as what the notion of just health care entailed. This involved maintaining or restoring mental and physical functions at or to normal levels, which was necessary to ensure fair equality of opportunity for all citizens. Insofar as this aim defines the goal of medicine, genetic enhancement falls outside this goal. Furthermore, insofar as this type of intervention is not part of the goal of medicine and has no place in a just health care system, there are no medical or moral reasons for genetically enhancing normal human functions and capacities.

Some have argued that it is mistaken to think that a clear line of demarcation can be drawn between treatment and enhancement, since certain forms of enhancement are employed to prevent disease. Leroy Walters and

Julie Gage Palmer refer to the immune system as an example to make this point:

> In current medical practice, the best example of a widely accepted health-related physical enhancement is immunization against infectious disease.
>
> With immunizations against diseases like polio and hepatitis B, what we are saying is in effect, "The immune system that we inherited from our parents may not be adequate to ward off certain viruses if we are exposed to them." Therefore, we will enhance the capabilities of our immune system by priming it to fight against these viruses.
>
> From the current practice of immunizations against particular diseases, it would seem to be only a small step to try to enhance the general function of the immune system by genetic means. . . . In our view, the genetic enhancement of the immune system would be morally justifiable if this kind of enhancement assisted in preventing disease and did not cause offsetting harms to the people treated by the technique.[26]

Nevertheless, because the goal of the technique would be to prevent disease, it would not, strictly speaking, be enhancement, at least not in terms of the definitions given at the outset of this section. Genetically intervening in the immune system as described by Walters and Palmer is a means of maintaining it in proper working order so that it will be better able to ward off pathogens posing a threat to the organism as a whole. Thus, it is misleading to call this intervention "enhancement." When we consider what is normal human functioning, we refer to the whole human organism consisting of immune, endocrine, nervous, cardiovascular, and other systems, not to these systems understood as isolated parts. The normal functioning in question here pertains to the ability of the immune system to protect the organism from infectious agents and thus ensure its survival. Any preventive genetic intervention in this system would be designed to maintain the normal functions of the organism, not to restore them or raise them above the norm. It would be neither therapy nor enhancement but instead a form of maintenance. Therefore, the alleged ambiguity surrounding what Walters and Palmer call "enhancing" the immune system does not impugn the distinction between treatment and enhancement.

If enhancement could make adequately functioning bodily systems function even better, then presumably there would be no limit to the extent to which bodily functions can be enhanced. Yet, beyond a certain point, heightened immune sensitivity to infectious agents can lead to an overly aggressive response, resulting in autoimmune disease that can damage healthy cells, tissues, and organs. In fact, there would be a limit to the beneficial effects of genetic intervention in the immune system, a limit beyond which the equilibrium between humoral and cellular response mechanisms would be disturbed.[27] If any intervention ensured that the equilibrium of the immune system was maintained in proper working order, then it would be inappropriate to consider it as a form of enhancement.

To further support the treatment-enhancement distinction, consider a nongenetic intervention, the use of a bisphosphonate such as alendronate sodium. Its purpose is to prevent postmenopausal women from developing osteoporosis, or to rebuild bone in women or men who already have osteoporosis. Some might claim that, because it can increase bone density, it is a form of enhancement. But its more general purpose is to prevent bone fractures and thus maintain proper bone function so that one can have normal mobility and avoid the morbidity resulting from fractures. In terms of the functioning of the entire organism, therefore, it would be more accurate to consider the use of bisphosphonates as prevention, treatment, or maintenance rather than enhancement.

Some might raise a different question. Suppose that the parents of a child much shorter than the norm for his age persuaded a physician to give him growth hormone injections in order to increase his height. Suppose further that the child's shortness was not due to an iatrogenic cause, such as radiation to treat a brain tumor. Would this be treatment or enhancement? The question that should be asked regarding this issue is not whether the child's height is normal for his age group. Rather, the question should be whether his condition implies something less than normal physical functioning, such that he would have fewer opportunities for achievement and a decent minimum level of well-being over his lifetime. Diminutive stature alone does not necessarily imply that one's functioning is or will be so limited as to restrict one's opportunities for achievement. Of course, being short might limit one's opportunities if one wanted to become a professional basketball player. But most of us are quite flexible when it comes to formulating and carrying out life plans. Robert Reich, the treasury secretary in President Clinton's first administration, is just one example of how one can achieve very much in life despite diminutive stature. If a child's stature significantly limited his functioning and opportunities, then growth-hormone injections should be considered therapeutic treatment. If his stature were not so limiting, then the injections should be considered enhancement.

Admittedly, there is gray area near the baseline of adequate functioning where it may be difficult to distinguish between treatment and enhancement. Accordingly, we should construe the baseline loosely or thickly enough to allow for some minor deviation above or below what would be considered normal functioning. An intervention for a condition near the baseline that would raise one's functioning clearly above the critical level should be considered an enhancement. An intervention for a condition making one's functioning fall clearly below the baseline, with the aim of raising one's functioning to the critical level, should be considered a treatment. For example, an athlete with a hemoglobin level slightly below the norm for people his age and mildly anemic may want to raise that level significantly in order to be more competitive in his sport. To the extent that his actual hemoglobin level does not interfere with his ordinary physical functioning, an intervention to significantly raise that level would be an instance of enhancement. In contrast, for a child who has severe thalassemia and severe anemia, with the risk of bone abnormalities and heart failure, an intervention to correct the disorder would be an instance of treatment.

The main moral concern about genetic enhancement of physical and mental traits is that it would give some people an unfair advantage over others with respect to competitive goods like beauty, sociability, and intelligence. Unlike the cognitively disabled individual considered earlier, we can assume that their mental states would not be so different and that they would retain their identity. Enhancement would be unfair because only those who could afford the technology would have access to it, and many people are financially worse off than others through no fault of their own. Insofar as the possession of these goods gives some people an advantage over others in careers, income, and social status, the competitive nature of these goods suggests that there would be no limit to the benefits that improvements to physical and mental capacities would yield to those fortunate enough to avail themselves of the technology. This is altogether different from the example of immune-system enhancement. There would be no diminishing marginal value in the degree of competitive advantage that one could have over others for the social goods in question and presumably no limit to the value of enhancing the physical and mental capacities that would give one this advantage. Not having access to the technology that could manipulate genetic traits in such a way as to enhance these capacities would put one at a competitive disadvantage relative to others who would have access to it.

Advancing an argument similar to the one used by those who reject the treatment-enhancement distinction, one might hold that competitive goods collapse the categorical distinction between correcting deficient capacities and improving normal ones. This is because competitive goods are continuous, coming in degrees, and therefore the capacities that enable one to achieve these goods cannot be thought of as either normal or deficient.[28] Nevertheless, to the extent that any form of genetic intervention is motivated by the medical and moral aim to enable people to have adequate mental and physical functioning and fair equality of opportunity for a decent minimum level of well-being, the goods in question are not *competitive* but *basic*. In other words, the aim of any medical intervention by genetic means is to make people better off than they were before by raising or restoring them to an absolute baseline of normal physical and mental functioning, not to make them comparatively better off than others. Competitive goods above the baseline may be continuous; but the basic goods that enable someone to reach or remain at the baseline are not. Given that these two types of goods are distinct, and that they result from the distinct aims and practices of enhancement and treatment, we can affirm that enhancement and treatment can and should be treated separately. We can uphold the claim that the purpose of any genetic intervention should be to treat people's abnormal functions and restore them to a normal level, not to enhance those functions that already are normal.

As I have mentioned, genetic enhancement that gave some people an advantage over others in possessing competitive goods would entail considerable unfairness. A likely scenario would be one in which parents paid to use expensive genetic technology to raise the cognitive ability or improve the physical beauty of their children. This would give them an advantage over other children with whom they would compete for education, careers, and income. Children of parents who could not afford to pay for the technology would be at a comparative disadvantage. Even if the goods in question fell above the normal functional baseline, one still could maintain that such an advantage would be unfair. It would depend on people's ability to pay, and inequalities in income are unfair to the extent that they result from some factors beyond people's control.

We could not appeal to the notion of a genetic lottery to resolve the problem of fairness regarding genetic enhancement. For, as I argued in the last section, such a lottery is better suited to meeting people's needs than their preferences, and enhancements correspond to people's preferences. Moreover, a lottery might only exacerbate the problem by reinforcing the perception of unfairness, depending on how losers in the lottery interpreted the fact that others won merely as a result of a random selection. One suggestion for resolving the fairness problem (short of banning the use of the technology altogether) would be to make genetic enhancement available to all. Of course, how this system could be financed is a question that admits of no easy answer. But the more important substantive point is that universal access to genetic enhancement would not be a solution. Indeed, the upshot of such access would provide a reason for prohibiting it.

Universal availability of genetic enhancement would mean that many competitive goods some people had over others would be canceled out collectively. The idea of a competitive advantage gradually would erode, and there would be more equality among people in their possession of goods. There would not be complete equality, however. Differing parental attitudes toward such goods as education could mean differences in the extent to which cognitive enhancement was utilized. Some parents would be more selective than others in sending their children to better schools or arranging for private tutors. So, there still would be some inequality in the general outcome of the enhancement. But quite apart from this, the process of neutralizing competitive goods could end up being self-defeating on a collective level.[29] More specifically, one probable side-effect of boosting children's mental capacity on a broad scale would be some brain damage resulting in cognitive and affective impairment in some of the children who received the genetic enhancement. The net social cost of using the technology would outweigh any social advantage of everyone using it. If no one is made better off than others in their possession of social goods, but some people are made worse off than they were before in terms of their mental functioning, then the net social disadvantage would provide a reason for prohibiting collective genetic enhancement.

There is another moral aspect of enhancement that should be considered. I have maintained that inequalities above the baseline of normal physical and mental functioning are of no great moral importance and may be neutral on the question of fairness. Although equality and fairness are closely related, one does not necessarily imply the other. Again, fairness pertains to meeting people's needs. Once these needs have been met, inequalities in the possession of goods relating to preferences are not so morally significant. Thus, if the idea of an absolute baseline implies that people's basic physical and mental needs have been met, and if people who are comparatively better or worse off than others all have functioning at or above the baseline, then any inequalities in functioning above this level should not matter very much morally. If this is plausible, then it seems to follow that there would be nothing unfair and hence nothing morally objectionable about enhancements that made some people better off than others above the baseline. Nevertheless, this could undermine our belief in the importance of the fundamental equality of all people, regardless of how well off they are in absolute terms. Equality is one of the social bases of self-respect, which is essential for social harmony and stability.[30] Allowing inequalities in access to and possession of competitive goods at any level of functioning or welfare might erode this basis and the ideas of harmony and stability that rest on it. Although it would be difficult to measure, this type of social cost resulting from genetic enhancement could constitute another reason for prohibiting it.

Yet, suppose that we could manipulate certain genes to enhance our noncompetitive virtuous traits, such as altruism, generosity, and compassion.[31] Surely, these would contribute to a stable, well-ordered society and preserve the principle of fair equality of opportunity. Nothing in this program would be incompatible with the goal of medicine as the prevention and treatment of disease. But it would threaten the individual autonomy essential to us as moral agents who can be candidates for praise and blame, punishment and reward. What confers moral worth on our actions, and indeed on ourselves as agents, is our capacity to cultivate certain dispositions leading to actions. This cultivation involves the exercise of practical reason and a process of critical self-reflection, whereby we modify, eliminate, or reinforce dispositions and thereby come to identity with them as our own. Autonomy consists precisely in this process of reflection and identification. It is the capacity for reflective self-control that enables us to take responsibility for our mental states and the actions that issue from them. Given the importance of autonomy, it would be preferable to have fewer virtuous dispositions that we can identify with as our own than to have more virtuous dispositions implanted in us through genetic enhancement. These would threaten to undermine our moral agency because they would derive from an external source.[32]Even if our genes could be manipulated in such a way that our behavior always conformed to an algorithm for the morally correct course of action in every situation, it is unlikely that we would want it. Most of us would rather make autonomous choices that turned out not to lead to the best courses of action. This is because of the intrinsic importance of autonomy and the moral growth and maturity that come with making our own choices under uncertainty. The dispositions with which we come to identify, imperfect as they may be, are what make us autonomous and responsible moral agents. Enhancing these mental states through artificial means external to our own exercise of practical reason and our own process of identification would undermine our autonomy by making them alien to us.

In sum, there are four reasons why genetic enhancement would be morally objectionable. First, it would give an unfair advantage to some people over others because some would be able to pay for expensive enhancement procedures while others would not. Second, if we tried to remedy the first problem by making genetic enhancement universally accessible, then it would be collectively self-defeating. Although much competitive unfairness at the individual level would be canceled out at the collective level, there would be the unacceptable social cost of some people suffering from adverse cognitive or emotional effects of the enhancement. Third, inequalities resulting from enhancements above the baseline of normal physical and mental functioning could threaten to undermine the conviction in the fundamental importance of equality as one of the bases of self-respect, and in turn social solidarity and stability. Fourth, enhancement of noncompetitive dispositions would threaten to undermine the autonomy and moral agency essential to us as persons.

Negative and Positive Eugenics: Is There a Slippery Slope?

The two forms of genetic intervention that I have been discussing in this chapter could be characterized as eugenics, defined as "the use of science applied to the qualitative and quantitative improvement of the human genome."[33] "Eugenics" is almost universally regarded as a dirty word, owing largely to its association with the evil practice of human experimentation in Nazi Germany and the widespread sterilization of certain groups of people in the United States and Canada, earlier in the twentieth century.[34] One cannot help but attribute some eugenic aspects to genethical questions about the number and sort of people who should exist. But there is a broader conception of eugenics (literally "good creation" in Greek) that need not have the repugnant connotation of improving the human species.

The purpose of terminating an embryo with a mutation that would cause a disease, or of giving gene therapy to someone with a disease, is not to improve the human genome or the human species but instead to prevent or treat disease in identifiable people. The purpose is not the impersonal one of increasing the quantity and quality of types of experiences, but instead the person-affecting one of preventing harm to and benefiting people who have or would have to experience the symptoms associated with severe disease. Accordingly, while retaining a broad genethical focus, we should distinguish between positive and negative eugenics. The first type is motivated by the perfectionist ideal of improving the human species, whereas the second is motivated by the beneficent ideal of health promotion through disease prevention and treatment. To the extent that the aim of gene therapy is to prevent or control disease, and that the aim of genetic enhancement is to improve people's already normal traits and capacities, these two forms of genetic intervention correspond to negative and positive forms of eugenics.

Forms of eugenics not involving direct genetic intervention have been practiced since antiquity. In Plato's *Republic* and *Laws*, for example, an ideal society would encourage "judicious matings," meaning that mating between members of the ruling and mercantile classes would be discouraged. Only those people most likely to produce the "best" offspring were encouraged to mate, especially within the ruling class. This example of positive eugenics is morally objectionable because it involves discrimination on the basis of social class. By the same token, however, many people today select mates with whom they believe they will have children with favorable physical and intellectual traits, giving them a competitive advantage over others for social goods. This is also a version of positive eugenics, even though it does not involve genetic intervention. Nor would most people acknowledge it as such.

Selecting a specific mate in order to have children with specific traits and capacities by itself is not morally objectionable to the extent that it is not part of a state-sponsored program, does not involve any coercion, and does not give an unfair advantage to some people over others in having children. Although this practice may seem objectionable to some because it is motivated to have children with more than just normal physical and mental functioning, it could be defended on moral grounds. For, in the natural process of reproduction, it cannot be predicted precisely which traits a child will have, given the parents' genetic profile alone. Epigenesis and the incomplete penetrance of genotypes largely account for this uncertainty. On the other hand, genetically intervening to produce a child with specific traits and capacities might be objectionable because it could largely shape the child's fate and cast doubt on her autonomy and responsibility for the good she achieves in her life.

An example of negative eugenics without genetic intervention would be if a person from one race married a person from another race with the intention of not passing on any deleterious mutations and diseases to offspring. According to the "consanguinity coefficient," the more similar the genotypes of two people are, the more likely they will produce offspring with deleterious genes causing or predisposing them to certain diseases. Conversely, the more different the two genotypes are, the less likely they will pass on deleterious genes to their offspring. Thus, if a Caucasian woman were to marry an Asian man, and both believed that their different genotypes meant a higher probability of passing on normal genes to their children, then they could be said to practice negative eugenics. They would want to ensure that any of their children did not have a high risk of inheriting mutations that would likely result in disease. Unlike the example of positive eugenics given above, there would be nothing morally objectionable about this practice because the parents' intention would be to reduce the risk of disease in their children and thereby prevent them from harm.

Similar reasoning applies to the case of a rabbi who advises a man and woman who are Orthodox Ashkenazi Jews and carriers of the Tay-Sachs allele not to marry, or else not to have children. Provided that the rabbi's advice was not coercive and the couple made a voluntary, informed decision, this too would be a morally defensible form of negative eugenics. It would aim at preventing harm by eliminating the risk of having a child with Tay-Sachs disease.

It might be more appropriate to call this last preventive strategy a form of *euphenics*, a beneficial manipulation of environmental factors to prevent or treat diseases.[35] Better education and nutrition would be more effective ways of achieving this goal in the general population. Still, genetic testing for the presence of mutations making people susceptible to various diseases can be part of a general euphenics program. Knowing the symptoms and genetic cause of a disease may enable us to devise treatments that can control its symptoms. Again, the best-known disorder fitting this description is PKU. Restricting an affected child's dietary intake of phenylalanine can neutralize the harmful effects of this disorder and ensure a life without severe mental retardation. In addition, people with sickle-cell anemia can avoid morbidity by taking penicillin, and those with alpha-thalassemia may be cured by bone-marrow transplantation. The social environment can play an important role as well. I already have pointed out that the severity of schizophrenia can be controlled to some extent by familial and social support of those who have the disorder. Despite differences of definition, ultimately the goal of both negative eugenics and euphenics is the same— health promotion through prevention and treatment of disease.

As in the distinction between gene therapy and genetic enhancement, the main distinction between negative and positive eugenics is that the first is based on the principle of beneficence and the second on the principle of perfectionism. In gene therapy, the goal is to benefit people by restoring or raising them to adequate physical and mental functioning and giving them opportunities to achieve a decent minimum level of lifetime well-being. In genetic enhancement, the goal is to give additional benefits to people who already have adequate functioning, perfecting their traits and capacities and giving them an advantage over others in competing for social goods. In the first case, we are morally obligated to create people *without* certain traits, or to remove these traits once they exist. But we are obligated to do this only if the traits in question cause severe disease and disability and severely restrict people's opportunities. In the second case, we are morally obligated to create people *with* certain traits, or to add them to existing people in order to raise their functioning and increase their opportunities above the norm for persons. But how can we be so sure that negative eugenics will not evolve into positive eugenics? Is there not a slippery slope here?

Bernard Williams notes that the slippery-slope argument "is often applied to matters of medical practice. If X is allowed, the argument goes, then there will be a natural progression to Y."[36] For present purposes, we can take X to represent gene therapy and negative eugenics, and Y to represent genetic enhancement and positive eugenics. The natural progression from X, with the ostensible aim of raising or restoring people's functioning to a normal level, to Y leads to the "horrible result" of positive eugenics at the bottom of the slope.[37] Presumably, what makes the eugenic slope slippery is that once we get on the negative side, we cannot get off and fall on to the positive side. The point of the argument is that we should not get on the slope to begin with. Negative eugenics is not morally justifiable and should not be practiced, because it inevitably leads to the positive eugenics and the violations of human value and dignity that it entails.

Many invoke the slippery-slope argument to reject controversial practices in biotechnology. But if the argument is to serve as a ground for rejecting these practices, then it must be sound. True premises must entail the conclusion. Let us now examine the logical form of the argument as it is applied to eugenics and determine whether it is sound.

There are three different species of the slippery-slope argument: conceptual—relating to vagueness of terms; precedential—relating to the need to treat similar cases consistently; and causal—relating to the avoidance of the actions that will initiate a sequence of events leading to an undesirable result.[38] The classic or generic argument, the one most often advanced in discussions of biotechnology, includes aspects of all three species. First, defenders of the slippery-slope argument exploit any vagueness in the definition of terms. They claim that the difference between treatment and enhancement is vague, since many enhancements really are treatments, and vice versa. Treatments and enhancements fall along a single continuum of medical interventions; the difference between them is one of degree rather than of kind. Second, they claim that, since treatments and enhancements involve only differences of degree, cases germane to one are assimilable to cases germane to the other. Consistency requires that we treat relevantly similar cases in the same way, and since cases of treatment and enhancement (negative and positive eugenics) are relevantly similar, cases of one are assimilable, or logically linked, to cases of the other. Third, they claim that, since treatment and enhancement are assimilable to each other, cases of one will cause cases of the other. This embodies the idea of the dangerous precedent. That is, case (a) may be prima facie acceptable, while cases (b), (c), and (n) are not. Yet, because (a) is relevantly similar and thus assimilable to (b), (c), and (n), doing (a) would set a dangerous precedent, as it would cause (b), (c), and (n). Therefore, (a) should not be permitted.[39]

Incorporating aspects of all three of the more particular arguments that I have just laid out, the logical form of the more general slippery-slope argument for the issue at hand looks something like this:

1. Case (a)—an instance of treatment, negative eugenics—is prima facie acceptable.
2. But cases (b), (c), . . . and (n)—instances of enhancement, positive eugenics—are unacceptable.
3. Cases (a) through (n) are assimilable, as they differ from each other only in degree, falling along a continuum of cases of the same type.
4. Case (a), if permitted, will be a precedent for cases (b) through (n).
5. Permitting (a) will cause (b) through (n).
6. Therefore, case (a) should not be permitted.

If we are to accept the conclusion of this argument, then it must be sound. The conclusion must follow from the premises, and premises (1) through (5) must all be true.

But the truth of (3) can be questioned. In fact, treatment and enhancement are different in kind, not merely degree, and they correspond to distinct aims that can be clearly articulated. If this is correct, then premise (3) is false, since cases of treatment are not relevantly similar and not assimilable to cases of enhancement. Premise (3) is false owing to the fallacy of assimilation. Furthermore, if case (a) is not relevantly similar to cases (b) through (n), then it is unlikely that (a) would cause (b) through (n) to occur. Hence premise (5) is false as well. Owing to the falsity of two of its premises, the argument is unsound. And because it is unsound, the claims of those who believe that negative eugenic treatments inevitably will lead to positive eugenic enhancements are ungrounded. At the very least, the burden of proof is on those making these claims to formulate a sound and more cogent argument to support their position.

Considerations of the truth or falsity of the argument aside, it seems plausible to say that adequate regulations and safeguards could prevent the alleged slide from occurring. There is no reason why public policy and legislation cannot guarantee this through a clearly spelled-out distinction between the aims and methods of treatment and enhancement and enforce the permissibility of the first and impermissibility of the second.[40] The negative eugenics that I am defending has affinities with what Philip Kitcher has called "utopian eugenics."[41] This consists of a policy guaranteeing reproductive freedom in choosing whether to bring people into existence, as well as access to medical technology to raise or restore physical and mental functioning to, or maintain it at, an adequate level. These choices must be free of any social or political coercion to prevent people from existing for economic reasons or perfectionist ideals. Provided that the technology of which we avail ourselves in making these choices is for the purpose of preventing or treating disease, rather than enhancing already adequate functioning, and that this technology is accessible to all, utopian eugenics can be a morally justifiable policy.

Conclusion

I have addressed two forms of genetic intervention in this chapter: gene therapy to treat diseases and restore or raise people's physical and mental functioning to a normal level; and genetic enhancement to raise people's functioning above this level. Gene therapy is motivated by the medical goal of health promotion through disease prevention and treatment and is grounded in the moral principles of nonmaleficence, beneficence, and justice. It is meant to prevent harm to and benefit people by preventing disease in them and providing them with opportunities to undertake and complete projects within a life plan of their own making. I explained how gene therapy at different stages of developing human organisms can affect the identities of the people who come into existence. In addition, I demonstrated how identity bears on our understanding of beneficence and justice in cases of cognitive gene therapy.

If we are to benefit the greatest number of people afflicted with different diseases, then there are good reasons to seriously consider shifting genetic research and treatment priority away from diseases like cystic fibrosis, Tay-Sachs, and Huntington's to heart disease, diabetes, and cancer. These reasons include the comparatively smaller number of people with severe but rare monogenic disorders, the larger number of people with more common multifactorial disorders, the extent to which people within these two classes benefit from treatment, and the fact that medical resources are scarce. Some might argue that we should shift emphasis away from genetic research in general to public health programs and prenatal care, which can prevent many chronic diseases in a significantly large number of people. Gene therapy has shown some promise in treating such disorders as severe combined immune deficiency (SCID), familial hypercholesterolemia, hemophilia, and alpha-thalassemia, but has not been proven effective for most diseases. Presently, the most effective course of action is not to try to treat severe genetic diseases but to prevent them from occurring. As I argued in chapter 2, this would involve genetic testing and selective termination of embryos with mutations that would have caused these diseases in the people we otherwise would have brought into existence.

I also considered the prospect of genetic enhancement and gave four reasons why it would be impermissible to improve physical and mental traits and capacities above the functional norm for people: It would involve unfair access to enhancement technology based on ability to pay; it would involve an unacceptable social cost in the form of mental impairment as a side-effect in some people; it would threaten to undermine equality as one of the bases of self-respect, social stability, and solidarity; and it would threaten to undermine individual autonomy and responsibility. I characterized treatment and enhancement as negative and positive forms of eugenics and argued that one does not inevitably lead to the other. Finally, I demonstrated that the slippery-slope argument advanced to support the claim that there is a slide from negative to positive eugenics is unsound owing to the fallacy of assimilation and an unsupported claim about causation between distinct types of cases in two premises of the argument.

Cloning is another form of intervention in human genesis that raises additional questions about genetics and eugenics. The molecular biology of cloning, the motivation for cloning human body parts, embryos, and even full-fledged human beings, and whether different forms of cloning are desirable and defensible on biological, medical, and especially moral grounds need to be discussed separately.

4

THE MORALITY OF
HUMAN CLONING

In 1997, Ian Wilmut and his colleagues announced that they had cloned a
viable lamb by transplanting the nucleus of a somatic cell from a six-year-
old sheep to an enucleated egg.[1] This finding raised the possibility of
cloning human cells to grow tissues and organs for transplantation. It also
raised the possibility of cloning complete human organisms. The fascina-
tion surrounding the prospect of cloning human beings has been tempered
with caution, however. On the recommendations of the National
Bioethics Advisory Commission, President Clinton drafted the Cloning
Prohibition Act of 1997, which outlawed somatic-cell nuclear transfer for
the purpose of creating a human being. Significantly, the Act did not call
for an outright ban on all cloning research, but included a five-year "sunset
clause" that allowed important and promising work to clone DNA, cells,
tissues, and nonhuman animals.[2]

One possible development of this research is therapeutic cloning, where
embryos are cloned from patients' own somatic cells. Because a cloned
embryo would contain a patient's own DNA, it could be produced to develop cells and tissues that would be compatible with the patient's immune
system and avoid graft rejection. Embryos could be cloned to derive embryonic stem (ES) cells, which could be used to rejuvenate cells, tissues,
and organs damaged from degenerative diseases like arthritis, diabetes, and
Parkinson's. Cloning embryos thus could play an important role in regenerative medicine. I will discuss this possibility later in this chapter. ES cells
also could be derived from embryos created through IVF that have been
discarded by fertility clinics. I will discuss the moral implications of this
particular source of ES cells in chapter 5. Therapeutic cloning is now legal
in Britain. Canada issued draft legislation in May 2001 permitting stem
cell research, but prohibiting human cloning and the creation of embryos
solely for research purposes. And in July 2001, The U.S. House of Repre-

sentatives voted by an overwhelming majority to ban cloning, not only for reproduction but also for medical research, even for therapeutic reasons.

The rationale for a prohibition on cloning humans is understandable. It is not yet known whether the technique would be safe for the individuals resulting from it. At a deeper level, cloning raises metaphysical and moral questions about what makes us human. Specifically, it has been argued that cloning is morally repugnant and should be banned because it involves a form of biological manipulation that violates fundamental features of our humanity.[3] For some, this technology evokes visions of Huxley's *Brave New World*, where cloning people into different groups is part of a comprehensive program of social engineering that deprives individuals of their freedom. Although cloning is in many respects an extension of existing procedures of assisted reproduction, such as artificial insemination and IVF, the asexual reproduction of genetically identical copies of ourselves seems anathema to our deep-seated conviction about the uniqueness and dignity of each person. For others, cloning does not threaten human dignity because dignity is a function of much more than the biological means of reproduction through which humans are conceived and caused to exist.[4] Furthermore, cloning could avoid many of the diseases that result from sexual reproduction by preempting what Joseph Fletcher has called "genetic roulette."[5] These and other related issues will be discussed in this chapter.

Possible uses of cloning include: (1) "replacing" a loved one who is dying or who has just died; (2) producing an individual with specially designed physical traits; (3) producing a child for an infertile couple, when all other reproductive alternatives have been exhausted; (4) producing a child for a lesbian or gay couple; (5) cloning either of two genetic parents, when each of them is a carrier of a mutant allele that in normal reproduction would entail a significant risk of having a child with a severe disease; (6) mining stem cells from cloned embryos to rejuvenate tissues damaged from disease in adults; (7) genetically reprogramming cloned preimplantation embryonic cells to prevent diseases by correcting mutations before the embryos develop into people; and (8) growing organs for transplantation.

Intuitively, these possibilities fall along a rough continuum of moral justification, with (6), (7), and (8) being the most justifiable and (1) and (2) the least. For while (6)–(8) are the most likely to prevent harm to persons by preventing, treating, or curing disease, (1) and (2) are the most likely to cause harm to persons by treating them solely as means and not also as ends in themselves. I will flesh out these intuitions by exploring the moral implications of each of the eight uses I have mentioned. After addressing the biological aspects of cloning, I will articulate the grounds on which (1)–(8) can or cannot be morally justified. In examining these issues, I will be guided mainly by the Kantian deontological injunction to "treat humanity, whether in your own person or in that of any other, always as an end and never as a means only."[6] Crucially, this injunction says that it *is* morally permissible to treat people as means, but not *solely* as means. I then go on to suggest that, if the aim of cloning is to promote health by preventing, treating, or curing disease in people, then the strongest arguments for cloning will pertain to (6)–(8). These three uses effectively sidestep most moral objections because the cells that are cloned and their products are nonessential parts of persons without any moral status of their own.

Claims about cloning violating the autonomy, uniqueness, and dignity of persons are largely unfounded. The main moral concerns about cloning hinge on the biological possibility of DNA damage to somatic-cell nuclei and the adverse effects that asexual reproduction might have on the diversity of the human gene pool. Cloning could lead to an increase in genetic mutations in individuals and the human species as a whole, making people vulnerable to premature aging and disease and thus harming them. Although it is unlikely that cloning would be practiced on a scale broad enough to increase the number of mutations in the human gene pool, I explore this idea in response to the claim that cloning would reduce the number of genetic abnormalities and birth defects in children.

Biological Factors

There are two cloning methods to produce genetically identical copies of human organisms: embryo splitting, and the nuclear transfer of somatic cells. Embryo splitting consists in separating undifferentiated embryonic cells, or blastomeres, at an early stage of a developing organism. Because they still are undifferentiated, each cell is totipotent and therefore capable of giving rise to a different embryo and a different human being.[7] This is the method of cloning envisioned in Huxley's novel. Cloning through nuclear transfer is the method used by Wilmut and his colleagues to produce "Dolly" and the one with which I will be concerned in this chapter. Briefly, this consists in transplanting or injecting the nucleus of an adult body cell into an enucleated egg (oocyte) of another animal. Convinced that past efforts to clone mammals had failed because the donor cell that supplied the nucleus and the recipient egg were at different stages of the cell cycle, Wilmut et al. cut the supply of nutrients and thereby induced the donor cell to go into a quiescent phase in which the cell stops dividing. In this way, they were able to prepare the donor nucleus so that it would be compatible with the egg cytoplasm. This also enabled the transmuted nuclei to become reprogrammed to create every other kind of cell. The cells became capable of retaining all the genetic material necessary to produce a complete organism, given that every totipotent cell carries a complete set of genes for an organism. The reprogramming that takes place in the two-way transfer of proteins between the nucleus and the cytoplasm effects the de-differentiation of cells and allows them to become totipotent.

The difficulty with which Wilmut et al. produced one healthy lamb is illustrated by the fact that they obtained this result after beginning their experiment with 434 sheep oocytes. Of these, only 277 adult nuclei were transferred successfully to enucleated oocytes. The success rate of only 1 out of 434 oocytes indicates that cloning is not only an inefficient but also an exceedingly complicated biological process involving many different causal factors. In the case of human cloning, hundreds of eggs would be required to produce one human organism, and this number appears to be more than what a woman is physically capable of donating. Given the enormous failure rate just mentioned, attempting to clone human beings would seem to involve hundreds of failed attempts and an unknown number of viable, but perhaps deformed, offspring.

In July 1998, however, T. Wakayama and other researchers at the University of Hawaii achieved a much higher success rate, cloning mice from enucleated oocytes injected with cumulus cell nueclei.[8] A high percentage of these oocytes were developed in vitro and, once transferred, 151 of the original 195 oocytes were activated three to six hours after injection and developed to term. The suggestion of this result is that, in mammals, nuclei from adult somatic cells introduced into enucleated oocytes are capable of supporting full development. It also suggests that cloning full-fledged human organisms may not be just a remote biological possibility but probable and perhaps even likely in the foreseeable future. Indeed, in late 1998 it was reported that a South Korean physician, Lee Bo Yon, had produced a four-cell human embryo from genetic material extracted from a thirty-year-old woman.[9] Although fertility experts point out that it would take numerous experiments to prove that cloning could produce a healthy normal human birth, it may be just a matter of time before cloning becomes an acceptable way to conceive children for infertile couples who cannot avail themselves of any other type of assisted reproduction.

There are two sets of biological and moral reasons against cloning human beings into existence. The first set pertains to the genetic age of the DNA of the donor-cell nucleus that is injected into the enucleated oocyte. According to one theory of cell aging, cells follow a preset genetic recipe and divide only a certain number of times before dying.[10] This limit on cell replication is a property of the nucleus and its DNA, rather than the cytoplasm of the cell. Among other mechanisms, the aging of cells is controlled by telomeres, the ends of chromosomes consisting of repeating sequences of DNA. Telomeres become increasingly shorter the more times cells divide until they no longer can protect other vital parts of the chromosome. At a certain point, the cell stops dividing and dies. Alternatively, there is a theory according to which cells age through random damage to DNA as it interacts with elements inside and outside the body.[11] These include elements in the external environment, such as chemicals in cigarette smoke, ionizing radiation, and pesticides, as well as chemicals inside cells, such as hydroxyl radicals, which are waste products of cell metabolism. These can cause mutations in the sequences of base pairs that constitute the DNA molecule. As a person ages, these mutations adversely affect the repair mechanisms in cells that ordinarily correct for this damage. On this second theory of aging, irreversible DNA damage accumulates as the organism ages. Consequently, proteins that control normal cell function are altered, telomeres are prematurely shortened, and there is an increased likelihood of premature aging and disease.

If the first theory of cell aging in correct, and the donor's biological age was not too advanced, then a cloned individual probably would develop normally and not age prematurely or suffer from any adverse health effects of genetic mutations in its cells. Assuming that there is compatibility between the donor-cell nucleus and the recipient egg cytoplasm, and that the donor-cell nucleus has been making the enzyme telomerase all along, normal telomerase activity and a normal rate of cell division would occur in the cloned individual. But if the donor-cell nuclear DNA has been exposed to radiation and carcinogens, then the clone's cells would be considerably older than its chronological age. It would age faster than the normal biological rate. Such an individual would be at increased risk of inheriting genetic mutations caused by cumulative damage in the donor-cell nuclear DNA and consequently would be more likely to experience premature aging and disease. In fact, geneticists have found that Dolly's genes began to show signs of wear not long after her birth. The udder cell used to create Dolly was from a six-year-old (very old for a sheep) with already shortened telomeres, and research has shown that Dolly's telomeres are 20 % shorter than those of sheep of her same age that were not cloned.[12]

These biological factors have significant moral implications for the cloning of human beings. Cloning could result in the existence of individuals who would experience premature disease due to genetic mutations in the donor-cell DNA. This would harm individuals by defeating their interest in a healthy existence and living out a reasonably disease-free life span. The possible biological effects of cloning on the aging process would indirectly cause harm in people by producing a diseased bodily condition that they would have to experience. This is one moral ground on which cloning humans would be impermissible.

Furthermore, if cloning were performed in vitro, then mutations might occur while somatic cells were growing in culture. Even here, exposure to certain environmental factors of the sort I have mentioned might cause genetic damage. In this scenario, it would be difficult to determine whether the donor-cell nucleus was normal or abnormal; there would be no way of ascertaining the extent of damage to its DNA due to factors inside or outside the cell. Thus there would be no way of knowing whether a cloned individual would develop normally or experience premature aging and disease. Here too the clone could inherit genetic damage that would defeat its interest, once it exists, in having a reasonably disease-free life span and thus would be harmed.

The issue is not whether a donor-cell nucleus is a potential person with a right to be brought into existence. Rather, the issue is that a cloned individual, given that she exists or will exist, has an interest in living out a life span without having to experience premature aging and disease. And given that harm consists in the defeat of a person's interests, a cloned person may be harmed by having to live with premature disease resulting from genetic damage to the donor cell nucleus that is transferred to the oocyte. To the extent that there is no way of knowing whether there would be genetic damage, that merely potential people do not have a right to be brought into existence by any means, and that the potential risks in cloning seem to outweigh the potential benefits, it seems plausible to say that we should remain on the side of caution and prevent this state of affairs from obtaining. This is the first set of biological and moral reasons for not cloning human beings into existence.

More generally, as an asexual form of reproduction, widespread cloning through nuclear transfer could result in serious long-term costs to genetic diversity and the survival capacity of the human species. Sexual reproduction allows for genetic variation among offspring, which minimizes the number of deleterious mutations and in turn enhances the ability of a species to adapt to and survive in changing physical environments.[13] In particular, genetic variation can help organisms evolve in such a way as to cope better with infectious agents. This confers a critical survival advantage on the human species, which enables us to transfer our genes into the next generation. Cloning interferes with the rate of adaptive evolution, which is much slower in asexual populations.[14] One reason for this is that two or more beneficial mutations that arise in different lineages can be combined into the same lineage through sex, an advantage that sexual populations have over asexual ones.

Silver claims that, in birth defects resulting from an abnormal number of chromosomes, such as Down syndrome, the abnormality occurs because of mistakes when genetic material is reduced by half in the process of egg formation. Cloning would greatly reduce the likelihood of this occurring, because the process of somatic-cell transfer involves no such reduction in genetic material.[15] Moreover, in recessive disorders such as CF, Tay-Sachs, SCA, and PKU, where the affected child inherits two copies of a mutant allele carried by each parent, cloning one of the parents would mean that only one defective allele would be transmitted to the cloned child. Strictly speaking, the clone would be the genetic sibling of the parent. More importantly, the child would be only a carrier of the trait and therefore would avoid having the disease. This corresponds to scenario (5) described earlier in this chapter. The general incidence of birth defects in cloned children would be lower than that in naturally conceived children.[16]

In the light of these points, one could argue that cloning would improve the human species by reducing the incidence of disease in people. Assuming that parents could choose cloning over sexual means of reproduction, individuals who came into existence through cloning in the present and near future might benefit from it. Yet in evolutionary terms the human species as a whole over many generations might be worse off if asexual forms of reproduction like cloning became widespread. Leon Eisenberg supports this position:

> Cloning would be a poor method indeed for improving on the human species.
> If widely accepted, it would have a devastating impact on the diversity of the

human gene pool. Cloning would select for traits that have been successful in the past but that will not necessarily be adaptive to an unpredictable future. Whatever phenotypes might be produced would be extremely vulnerable to the uncontrollable vicissitudes of the environment.[17]

With respect to the survival capacity of the human species over many generations, cloning would be at odds with evolutionary biology. There are moral implications of this biological point. If cloning diminished the diversity of the gene pool and consequently our ability as a species to adapt to a changing physical environment, then persons in this and near future generations may benefit in the respects Silver highlights. But people in distant future generations might be harmed by their inability to cope with infectious agents, among other threats to them, in the environment. Potential present benefits to some must be weighed against potential future harms to others. If the number of people who could be harmed in the distant future is much greater than the number of people who stand to benefit from being cloned in the present and near future, and if we have a greater duty to avoid harm to a larger group of people than to benefit a smaller group, then it seems to follow that the interests of future people not to be harmed outweigh the interests of present people to benefit from cloning. This argument has to be advanced tentatively, however, since it is difficult to determine what the probability of harm would be to future people if cloning became a regular practice. In general, though, the claims of future people not to be harmed have as much moral weight as the claims of present people to be benefited. What matters is not *when* they exist, but *that* they will exist. We cannot discount their claims simply because they do not yet exist and only will exist in the distant future. Existence, rather than time, is what makes present and future people equal in both their metaphysical and moral status.[18]

Sexual reproduction also confers a survival advantage on a species by enhancing DNA repair mechanisms in organisms belonging to that species. This results from the process of "outcrossing," which consists in the fusion of two cells in sexual reproduction so that the genetic material of each parent cell becomes enclosed inside a single membrane.[19] Enzymes are then produced that can check for and repair damage to sequences of chromosomal base pairs. In this way, the outcrossing in sexual reproduction can prevent the occurrence and expression of genetic mutations. Over many generations, cloning, as an asexual form of reproduction, could result in the expression of more deleterious mutations and a higher incidence of

premature aging, disease, and death in humans. It could threaten the natural variation necessary for one generation to survive long enough to pass
its genes on to the next and over time may adversely affect the ability of
the human species to survive. This is the second set of biological and moral
reasons against cloning.

Some have claimed that cloning on a broad scale is so unlikely that its
effects on genetic diversity are not worthy of serious consideration. For
example, Dan Brock says that it "is not a realistic concern since human
cloning would not be used on a wide enough scale, substantially replacing sexual reproduction, to have the feared effect on the gene pool. The
vast majority of humans seem quite satisfied with sexual means of reproduction."[20] But it is not so fanciful to imagine a substantial number of
prospective parents who might want to avoid playing the game of genetic roulette and opt for cloning. If both parents knew that they were
carriers of a mutant allele that could cause a recessive disorder, then they
might choose cloning instead of sexual reproduction to eliminate the
risk of having a diseased child. Furthermore, if prospective parents believed that cloning could eliminate mutant alleles that increased lifetime
susceptibility to moderately severe conditions like osteoarthritis and cardiovascular disease, then a number of them might choose cloning over
sexual reproduction to have children who would not be susceptible. This
scenario may be less rather than more likely to occur over time. Still, the
point of describing it is to illustrate how cloning could increase the
number of genetic mutations in humans in the long term . And even if it
did not adversely affect the human species as a whole, it could increase
the number of mutations within families that reproduced through
cloning over many generations.

In this section, I have addressed some of the possible long-term biological risks of human cloning and have explored the moral implications of
these risks for future people. This is not to suggest that all forms of cloning
should be banned. On the contrary, there are compelling medical reasons
for certain forms of cloning in particular cases that may be of great benefit
to people. In the next section, I will examine these forms and the effects
they might have on people, paying particular attention to the question of
whether they are morally permissible and on what grounds they would be
permissible. More specifically, I will evaluate them in terms of whether
they are consistent with the Kantian injunction to treat persons not solely
as means but also as ends in themselves. This will be construed broadly
enough to include the question of how cloning affects people's dignity and

the autonomy to make choices and act in accord with a life plan of their own making.

Means and Ends

Earlier, I cited eight possible forms of cloning, though only three of these pertained to the production of full-fledged human organisms. The others pertained to embryos, tissues, organs, and other human body parts. Let us examine each of these two general forms of cloning in more detail.

Suppose that the parents of a recently deceased or dying child want to clone an individual who is genetically identical to that child and thus "replace" it to compensate for their loss or else carry on the family line. If the sole intent of the parents is replacement or compensation, then the cloned individual would be treated solely as a means, which would violate the second formulation of Kant's categorical imperative. A person is not a mere extension of its parents or siblings, and to treat one as such would deny the intrinsic dignity and worth one possesses in virtue of the fact that one is a human agent with the capacity for reason.[21] A parent also may want to create a child with the same physical features that he has, which would be the highest form of egoism and narcissism on the parent's part. Yet, despite the apparent repugnance of creating another individual in one's own image, or to fill a void left by a deceased child, if the clone were loved and treated with the dignity and respect commanded by its intrinsic worth, then cloning might be morally justifiable on Kantian grounds. Although the intention to clone the child suggests that he or she would be treated instrumentally, the fact that the child is treated as a unique individual once she exists is enough to dispel any moral qualms about the parent's behavior. What matters is how people are treated over the course of their lives as a whole, not the intention for causing them to exist. Besides, parents have children for a variety of reasons, some of which may seem questionable, and there is no justification for treating the reason for having a cloned child as a special case.

Parents may decide to clone a brother from a son dying from a blood disorder like leukemia. Because the cloned brother would be genetically identical to his sibling, his bone marrow would be compatible with that of his sick brother and would not be rejected by his immune system. As soon as the cloned brother's bone marrow was mature enough for harvesting, it could be extracted and transplanted into the older brother to cure him of his disease and save his life. This recalls the case of Anissa Ayala of Walnut

Grove, California, in the late 1980s.[22] While Anissa was dying slowly from leukemia, her parents decided to have another child (a daughter) on the chance that she would be a donor match with Anissa. Fortunately, the bone marrow of the younger daughter, Marissa, was compatible with Anissa's, the marrow was transplanted when it was mature enough, and Anissa's life was saved. Although at first blush the parents' decision and action seemed morally unjustifiable, all accounts now suggest that the child who was conceived in order to save her sister's life has been loved and respected by her parents and sister as a distinct individual with her own intrinsic worth. They have treated her with dignity by respecting the fact that she has a life of her own. In other words, the Ayalas *were* treating the younger daughter as a means. Yet they have not treated her *solely* as a means, but *also* as an end in itself.

In this and the other case that I have presented, cloning a genetically identical but numerically and qualitatively distinct individual to save or replace another does not by itself imply that the clone is treated solely as a means. If he also is treated as an end in itself, then the act of cloning a human being is not inherently morally objectionable. For, on the Kantian view, treating a person as such implies respect for that person as an autonomous locus of dignity and worth, one whose capacity for reason enables him to have interests and rights of his own. Insofar as respecting a person means upholding his interests and rights, and insofar as cloning does not thwart his interests or violate his rights, a person is not harmed or wronged just because he has been cloned.

Suppose that, instead of conceiving another daughter through normal sexual reproduction and risking the odds that Marissa's bone marrow would not be compatible with Anissa's, the Ayalas cloned one of Anissa's somatic cells and were able to construct an embryo whose cells had the same genetic material as the original cell. The child born from the embryo would be genetically identical to Anissa and thus would be a perfect donor match. Again, her bone marrow would be transplanted into her sister, who would not have to worry about graft rejection due to an adverse immune response. Because of the genetic identity between the donor and the recipient, the transplant would have a higher chance of success. The same reasoning applies in this hypothetical scenario as in the actual one. Even if the donor were cloned from one of her sister's cells and were a genetically identical twin of Anissa, there would be nothing morally objectionable about the procedure and its outcome if the parents treated the clone respectfully as an individual with intrinsic worth over the course of its life.

In a similar case, Adam Nash was born in August 2000 from an embryo selected specifically because its tissue matched that of his six-year-old sister Molly, who had a deadly blood disorder.[23] The parents selected the embryo and allowed it to develop into a son so that stem cells could be harvested from his cord blood and transfused into Molly, thereby creating a new blood supply and immune system in the recipient. They had a second child in order to benefit a first child, and in this regard they used Adam as a means. As in the case of Marissa Ayala, however, Adam's parents have treated him with love and respect as a unique person. From the fact that they treated their child as a means in causing him to exist, it does not follow that they have treated him solely as a means. Insofar as they have treated him as an end in itself, one could not morally object to the method or reason for bringing him into existence. These cases involve the distinction between the natural and deliberate production of offspring. But this distinction should not matter morally. For, as noted earlier, parents have children for a variety of reasons, and it is difficult to distinguish justifiable from unjustifiable reasons for procreation. Also, how one is treated by others over the course of one's life is more morally significant than the reasons for causing one to exist.

George Annas writes that "the *only* reason to clone an existing human is to create a genetic replica. Using the bodies of children to replicate them encourages all of us to devalue children and treat them as interchangeable commodities."[24] But my description of the actual and hypothetical Ayala cases, as well as the Nash case, show that both of Annas's claims are false. In most imaginable cases, there would be complex reasons for cloning. If a child needed a lifesaving organ transplant, parents might create a genetic replica through cloning so that the genetic identity between siblings ensured that the transplant would not be rejected and the diseased child would survive. Moreover, insofar as the cloned child was treated as an end as well as a means by his parents and other relevant parties, cloning would not devalue the worth or undermine the dignity of the child.

Another scenario in which cloning might be medically and morally justified is one where a couple is infertile, due either to immature sperm or a low sperm count in the male, or damaged fallopian tubes in the female. Even here, though, cloning would be justified only if the couple had exhausted all other forms of assisted reproduction and did not undergo the procedure solely to have a child with particular traits. Lesbian and gay couples also might decide to have a cloned child so that it would be genetically related to at least one of them. One partner would provide the nucleus, the other the mitochondrial DNA and cytoplasm. Yet one would have to advance a persuasive argument for the intrinsic value of genetic relatedness between parent and child to justify creating a child through cloning for this purpose.

Some might hold that we naturally want to have our own genetically related children owing to a "primeval instinct" programmed into our genes.[25] But this idea implausibly suggests some form of genetic determinism, as though we were compelled by our genes to have the desire for children who are genetically related to us. It is implausible because, as I argued in chapter 1, genes alone do not determine the desires, intentions, and other mental states that make us persons. Who we are as persons, including the particular desires we have, is not simply a function of our genes. Moreover, there are five types of parenthood, only two of which are genetic. Owing to advances in reproductive technology, it is now possible for a child's genetic mother who contributes the egg to be distinct from the gestational mother, who in turn may be distinct from the social mother who actually rears the child. In fact, if a child has developed from an IVF embryo created through ooplasmic transplantation, then it would have two genetic mothers, one contributing the nuclear DNA, the other the mitochondrial DNA. Also, the child's genetic father who contributes the sperm may be distinct from the social father who rears the child. The special relation of emotional intimacy between parent and child is much more a function of social than genetic parenthood. It may be that this emotional relation depends on the other types of parenthood as well. But the genetic link cannot explain this on its own.

Alternatively, others might invoke the idea that procreation is a constitutionally protected right falling under the constitutional rights of privacy and liberty. Proponents of this view could cite the ruling of the U.S. Supreme Court, as stated in *Eisenstadt v. Baird*: "[I]f the right to privacy means anything, it is the right of the *individual*, married or single, to be free from unwarranted intrusion into matters so fundamentally affecting a person as the decision whether to bear or beget a child."[26] Presumably, this ruling means that a woman or couple would be entitled to have a genetically related child through nuclear transfer cloning, especially if all other forms of assisted reproduction had failed. Access to this reproductive technology could be justified by the principle of equal protection under the law. That is, because a woman may be infertile through no fault of her own and thus unequal to other women who are not infertile, access to cloning technology would be an equalizing factor for her in allowing her to exercise her right to have a genetically related child.

Curiously, the right to privacy and the phrase "freedom from" in *Eisenstadt* imply a negative right to bodily noninterference. Yet the right to bring a child into existence should be construed as a positive right and as such a claim by a person to receive aid or support. Although a violation of a negative right always is considered wrong, the same does not always hold for positive rights. People are not always wronged if their claims to receive aid or support are not met.[27] If bringing a genetically related child into existence is a positive rather than negative right, then it is not obvious that not having access to cloning technology would constitute a wrong to a woman or couple who wanted to exercise this right.

Furthermore, to the extent that they are claims on others, rights entail responsibilities on the part of persons making these claims for those on whom the claims are made. If a parent has a right to reproduce by any means, then she has a corresponding responsibility for the welfare of the child she brings into existence. This could mean *not* having a child if that child would have a life with severe disease or disability, defeating his interest, once he exists, in not having to experience these conditions. Reproduction involves the interests of at least two parties, the woman or couple who undergo the process, and the individual who comes into existence as a result of it. So, even if there is a right to reproduce, it is not absolute but only a prima facie right that may be overridden, depending on the health of the child who would be born. In addition, the right to reproduce may justifiably be overridden in a country with severe overpopulation, provided of course that the restriction on births did not involve discrimination on grounds of race, sex, or social class. In an overpopulated country with many unwanted children, an adoption program might ameliorate the problem, while continued reproduction would only exacerbate it. Even in North America, which does not have the same population problem as other parts of the world, there are enough unwanted children to question the right to reproduce by cloning or any other means.

Perhaps more importantly, the people who come into existence as a consequence of someone exercising their right to reproduce cannot consent to be brought into existence. They cannot consent to the act that causes them to exist or to the burdens that come with living through the different stages of their lives. If reproduction is a right, then it is a claim on others. Ordinarily, those on whom claims are made can consent or otherwise respond to these claims. But since people created by the reproductive process do not exist until the process is complete and therefore cannot respond when the right to reproduce is exercised, one can question whether there is a fundamental right to reproduce. If there is no such right, then no one would be entitled to have access to nuclear transfer cloning technology, legal principle and precedent notwithstanding.

On the question of how cloning would affect personhood and our humanity, Annas argues thus:

> Cloning would also radically alter what it means to be human by replicating a living or dead human being asexually to produce a person with a single parent. The danger is that through human cloning we will lose something vital to our humanity, the uniqueness (and therefore the value and dignity) of every human. Cloning represents the height of genetic reductionism and genetic determinism.[28]

Again, we should take issue with Annas's claims. Leaving aside the metaphysical distinction between persons and human beings that I drew and defended in chapter 1, let us go along with Annas and use "person" and "human" interchangeably. Creating a genetically identical clone of a parent or sibling does not amount to genetic reductionism or determinism because personal identity, or our humanity, is not equivalent to genetic identity. Cloning by itself does not threaten the uniqueness, value, and dignity of persons.

Having the same DNA as a parent or sibling may give a cloned child the same general physical features as that parent. As I have argued, however, personal identity consists in the persistence through time of one and the same individual identified with a body, brain, and a set of mental states including desires, beliefs, intentions, and memories unified from the conscious present. The nature and content of these states is shaped as much by, if not more so than, one's social and physical environment as they are by one's genetic makeup. Also, the ways in which the genetic material in the donor-cell nucleus expresses itself in the clone's phenotypic traits will be a function of the epigenetic process of development from the embryo to further stages of development of the human organism. This involves interactions between and among different stages of cell differentiation, as well as the interaction between the embryo or fetus and the uterine environment during gestation. There are different probabilities of genotypic penetrance in different people. Genes that are expressed as identifiable physical traits in the donor might not be expressed in the clone. Simply having the same genes as the person from whose cell one is cloned does not mean that the clone will be identical to its donor in physical or other relevant respects.

The genes that a cloned individual inherits from a parent or sibling may very well influence the psychological properties in which personal identity consists by shaping the physical properties of the body and brain that generate and sustain one's psychology. Still, genes alone cannot account for the distinctiveness of one's cognitive and emotional experience of and response to social and physical environments and therefore cannot completely determine the nature and content of the mental states definitive of personhood and personal identity. Genes influence who we are only within these environments. So it is unlikely that genes could be the primary explanation of one's psychological properties, owing to the complexity in the ways that environmental factors play a causal role in the etiology of one's cognitive and emotional states. Nor can genes account for the phenomenological quality of being conscious of having these states. Finally, the fact that parent and cloned child would be of different generations and therefore develop their psychology in different social environments means that their mental states would be qualitatively distinct.

At least one empirical study of the psychological traits of genetically identical twins supports these claims. In a questionnaire devised by psychologist Thomas Bouchard, there was a 50% correlation between the personality traits of genetically identical twins, based on different responses to the same questions.[29] This suggests that at least half of one's psychological traits are due to environmental factors operating independently of or in conjunction with genes. The more general upshot is that genetic identity underdetermines the sense of self that develops through one's experience in social and physical environments and one's construction of a unified life plan with goals and projects that confer meaning on one's life. In Bouchard's words, "selves, unlike cells, can never be cloned."[30] Ultimately, personhood and personal identity are not solely or even primarily functions of one's genetic ancestry, regardless of whether one came into existence through cloning or normal sexual reproductive means. These considerations should dispel any worries about genetic reductionism or genetic determinism.

A related concern about cloning human beings is that it threatens to undermine a person's freedom, her autonomy in forming and undertaking projects within a life plan of her own making. But however egocentric or otherwise misguided parents' aims may be in wanting to produce a clone of themselves, ensuring genetic identity through cloning would not necessarily restrict or undermine a child's freedom to develop a unique, inviolable self. What restricts or undermines a child's freedom, rather, is parents' refusal or inability to allow the child to develop her own life plan and values by imposing their own plan and values on her in an attempt to control her life. By not allowing the child to form and pursue her own conception of a good life, parents violate what Joel Feinberg has called "a child's right to an open future."[31] This phenomenon occurs all too often in many families. But surely it would not be confined to those with cloned children, because whether parents respect or deny this right to their children is not a function of genetic relatedness between parent and child.

The reason why concerns about cloning undermining human dignity are misguided is that they focus on the biological means through which humans come into existence. We possess dignity in virtue of being persons, not simply human beings. Since persons are psychological rather than biological kinds, dignity is a function of our psychology, of our autonomous desires, beliefs, and intentions, rather than a function of our biology. To be sure, biology plays a necessary role in our mental life insofar as structures and functions of our bodies and brains generate and sustain our psychology. But biology cannot completely account for all the features of our psychology. Even granting the general importance of biology for personhood, the particular biological fact of the reproductive means through which one comes into existence is of no real metaphysical or moral significance. Whether one is conceived and comes into existence through an asexual form of reproduction like cloning or through sexual reproduction has little to do with what makes one a person worthy of self-respect and respect from others.

Only if the production of genetic identity between parent or child (who, to repeat, would be the genetic sibling of the parent) through cloning were an integral part of the parent's plan to control the child's life would cloning be morally objectionable. It would be objectionable on the ground that it would violate the child's autonomy. By themselves, though, cloning and genetic identity are morally neutral; no moral value attaches to them. After all, we make no moral judgments about the fact that genetically identical twins result from the monozygotic twinning of an embryo in the natural unassisted form of human reproduction. This is largely because the twins go on to develop as numerically and qualitatively distinct individuals in virtue of developing distinct psychological properties over time. Yet, in the case of genetically identical twins, one is effectively a clone of the other. There is nothing inherently immoral or morally objectionable about the genetic identity entailed by cloning. What is immoral is the use of cloning to produce offspring who would be treated as mere means to one of the objectionable ends that I have described.

Reiterating the main points in this and the preceding section, there are three instances in which cloning would be morally objectionable: (1) if it entailed genetic damage to the donor-cell nucleus and resulted in premature disease in a cloned human being; (2) if, as an asexual form of reproduction practiced on a broad scale over many generations, it adversely affected genetic diversity, increasing the number of mutations in humans and limiting their ability to adapt to and survive in changing physical environments; and (3) if it meant treating a cloned individual solely as a means and not also as an end in itself. All of these instances would be morally objectionable because they would involve harm to people. But none of them derives from ungrounded and misguided worries about genetic determinism or genetic reductionism threatening the uniqueness of humans.

Cloning Body Parts

It is worth emphasizing that, when the National Bioethics Advisory Commission recommended a ban on human cloning in 1997, it also recommended that cloning of human cells be allowed for the purpose of producing tissues and organs. This is significant because, by cloning these human body parts, which are not identical to persons and have no intrinsic moral status of their own, we effectively sidestep any disturbing moral aspects of cloning full-fledged human beings. The most compelling reason for cloning of any sort is therapeutic. In the case of producing a genetic replica of an existing person from one of its somatic cells, the purpose may be to ensure that there will be a genetically identical donor for a transplant to cure that person of a life-threatening disease. Yet the same result could be achieved by cloning specific cell and tissue types from affected persons in order to create tissues and organs that could cure these same people of these same diseases. If this type of cloning were to become feasible, then there would be no biological or medical need to clone humans. Consistent with the point I made in chapter 2, the purpose of cloning would not be to create more people, but instead to improve or maintain the health of already existing people.

It would be difficult to find anything morally problematic about cloning human cells for the therapeutic purpose of controlling or curing disease in people. If this procedure were problematic, then it would either be because it resulted in some form of physical, cognitive, or emotional harm to people, or because it undermined personhood and personal identity. With respect to the first point, if cloning cells or other body parts is by definition therapeutic and can control or cure disease without any harmful side effects, then the procedure could only be beneficial to people. With respect to the second point, human body parts are not identical to persons. Although persons are constituted by cells and other body parts, they are not identical to these parts. Persons *have* body parts; but they *are* not body parts and cannot be described or explained entirely in bodily terms. Personhood and personal identity consist in more than the physical constituents of bodies. Furthermore, at the biological level of cells, tissues, and organs, there are no individuals with the mental capacity for interests and rights in virtue of which they might be harmed. On neither point, then, is there anything morally objectionable about cloning human body parts.

Yet, it is because of the causal dependence of our psychology on our biology that cloning body parts can benefit us as persons, given our interest in not experiencing premature aging and disease. In theory, researchers might be able to remove a small amount of a particular type of tissue from a patient, de-differentiate the cells in culture, and then genetically reprogram the cells to differentiate into a specific kind of cell for a particular organ that could be recognized by the immune system and not be rejected by it. This type of regenerative medicine may be realized sooner rather than later. Researchers have learned how to cultivate human embryonic stem cells, which are undifferentiated and totipotent, to construct custom-made neo-organs.[32] Ideally, we would want new tissues and organs that would replace diseased ones to grow within the person rather than within an artificial environment. But presently the most feasible way of doing this is to grow the cells in culture and then transfer the manufactured tissues or organs to the affected areas of the body. The cells would be transferred from the culture medium, in the form of a three-dimensional matrix, to the site where the desired tissue growth would unfold.[33] Some success already has been achieved in this regard with the manufacture of artificial skin and cartilage.

It is instructive to explore the therapeutic promise of the sort of genetic technology I have been discussing in this section. Consider polycystic kidney disease, in which cysts form in both kidneys and gradually become enlarged, destroying normal tissue and essential functions of this organ. The genetic defect that causes this disease may be either dominant or recessive. Those with a dominant mode of inheritance usually have no symptoms until adulthood; but those with a recessive mode of inheritance have severe illness in childhood. Suppose that an adult male who is an only child develops the disease. Because his mother has the disease and his father is diabetic, neither can donate a kidney to their son. In such a case, tissue could be removed from one of his kidneys and cultured cells from the tissue could be genetically reprogrammed to replace the genetically defective cells causing the disease. The same could be done for a younger child with the recessive genetic defect causing the disease if neither parent is able or willing to donate a kidney to him. This would avoid having to clone another human being who may or may not be a compatible kidney donor and would avoid the moral question of whether the clone was being used as a mere means.

Furthermore, in the case of someone suffering from leukemia, genetically defective cells causing the disease could be extracted and genetically reprogrammed in culture. They then could be injected back into the affected individual and produce normal amounts of white blood cells. Here too, this would obviate the need to bring another human being into existence for a therapeutic purpose and would preempt the moral question of whether such a person would be treated solely as a means. It would avoid any of the moral questions generated by the hypothetical cloning scenario raised as a variant of the Ayala case. Because the procedure would involve an autologous transplant, it would minimize the likelihood of rejection by the recipient's immune system.

Still another possibility of the therapeutic use of cloning cells to produce tissues and organs might involve diseased livers. The technique of autologous hepatocyte transplantation uses the patient's own liver cells. These cells, from portions of liver tissue diseased by one or more defective genes, could be removed and genetically modified in culture to repair the critical genes. They then could be infused back into the patient's liver, merging with the organ and replacing the defective hepatocytes.[34]

In many cases, however, patients suffering from leukemia, kidney disease, or liver failure may be too sick to benefit from such a procedure. Or, the damage to their cells may be so extensive that they could not be genetically reprogrammed. At an earlier time, before the onset of symptoms, cells could be stored for future production of replacement tissues or organs. Of course, this would assume that they knew they were likely to develop the disease, with the genetic information readily available to them. This is not to deny that we should proceed cautiously with research in this area, given the risk of DNA damage to cells stored for long periods of time. Nevertheless, when large numbers of people needing transplants die because there are not enough human donor organs, cloning organs would be one way of meeting this need.

In late 1997, molecular biologists led by Jonathan Slack created a frog embryo without a head.[35] They achieved this result by manipulating certain genes in such a way as to suppress development of a tadpole's head, trunk, and tail. This leads one to speculate whether the technique could be adapted to grow human organs such as hearts, kidneys, and livers in an embryonic sac functioning in an artificial womb. If such a technique were to become a reality, then people needing organ transplants could have their own custom-made organs grown from their own cloned cells. The embryos resulting from these cells could be genetically reprogrammed to suppress development of all parts of the body into which the embryo otherwise would develop, except for the needed parts, plus a heart and blood circulation. This also could alleviate the shortage of organs for transplantation. Even if one argued that embryos have moral status and that their further development should not be interfered with in any way, growing partial embryos could bypass legal restrictions and moral concerns. For without the potential for a central nervous system, these organisms would not technically qualify as embryos. Insofar as human body parts like cells and tissues lack moral status, genetically manipulating them in the respects I have described would not be morally objectionable. It would be objectionable only if it resulted in harmful effects on persons in the form of disease or disability over the course of their lives.

In addition, cloning human cells could be useful in correcting genetic mutations in early embryonic cells growing in culture. Consider sickle-cell anemia, which is caused by a mutation in an allele of the gene coding for hemoglobin. If genetic testing showed that embryonic cells had this mutation, then a normal functioning gene that would ensure proper functioning of hemoglobin could be inserted into embryonic cells by means of a vector. The DNA of the nucleus of one of the cells then could be implanted into a new enucleated egg from the mother and a new pregnancy could begin. In this way, the original embryo would be replaced by a healthy clone of itself, which would develop into a healthy human being.[36] Although the new embryo would be genetically identical to the original, the individual developing from the genetically normal embryo would go on to develop a different set of psychological properties and thus would be a distinct person. Still, the main issue here is the motivation for cloning cells or embryos, which is to prevent disease in people by intervening genetically before people come into existence. The aim is to prevent harmful effects in people by eliminating the genetic causes of diseases at an early stage of embryonic development. It is both medically and morally preferable to prevent diseases from occurring in the first place than to bring people into existence with severe diseases and then try to treat them, especially when treatments are not very effective.

Eugenics Again?

I have been arguing that in general cloning should be used, not to create more people, but to make existing people healthy. It would do this by preventing, controlling, or curing these diseases or by regenerating tissues and organs damaged by degenerative diseases. To the extent that it is both medically and morally more acceptable to achieve this aim by cloning body parts rather than full-fledged human beings, there are stronger reasons for supporting research on the first type and weaker reasons for supporting research on the second. Even here, though, one might ask whether at a deeper level the cloning of cells, tissues, and organs would be motivated by the desire to improve the human species and thus serve eugenic goals. If so, then would there not be moral grounds for banning all forms of cloning?

Before responding to this point, it is important to distinguish it from Eisenberg's point about the long-term evolutionary impact of cloning. The idea that cloning might be used to improve the human species conflicts with Eisenberg's claim that it would have precisely the opposite effect. It would reduce genetic diversity and thereby adversely affect the species' ability to adapt to and survive in changing physical environments. Yet Eisenberg's claim pertains explicitly to the effects on the human species and implicitly to the effects on individual people who will exist in the distant future. In contrast, claims about eugenics pertain to people who already exist or will exist in the near future. This is not to diminish the significance of the evolutionary impact of cloning. But the point about the eugenics of cloning is generated by more immediate concern about the welfare of people who already exist or will exist in the near future.

There are no moral grounds for banning all forms of cloning if we restate the distinction between positive and negative eugenics drawn and discussed in chapter 3 and place cloning in the second category. That is, if we use cloning to promote health by preventing or controlling disease, rather than to raise physical and mental capacities above the baseline of normal functioning, then there is nothing morally objectionable about the procedure. Reflecting on the alleged eugenic aspects of cloning, Leon Kass says that "we do indeed already practice negative eugenic selection, through genetic screening and prenatal diagnosis. Yet our practices are governed by a norm of health."[37] Kass goes on to differentiate health promotion from genetic enhancement, suggesting that while the latter is morally unjustifiable, the former is not only justifiable but obligatory. However, he also suggests that the line between normal physical and mental functioning and physical and mental enhancement cannot always be clearly drawn. The difference between the two is only of degree, which is one reason why cloning should be banned.

How can we be so sure that a cloning program intended as a form of negative eugenics would not evolve into a program of positive eugenics? Because the different forms of cloning would be arranged along a spectrum of cases of the same kind, if one case were permitted, then it could set a precedent for and cause other cases to be permitted. And because these other instances would involve morally repugnant "Brave New World" scenarios, cloning should not be permitted in the first place. Or so one might argue. It appears that the specter of the slippery slope once again has reared its ugly head.

Yet, as I demonstrated in chapter 3, the slippery-slope argument advanced in response to various uses of biotechnology is unsound because it rests on the fallacy of assimilation and on unwarranted claims about causation. Although the same general term is used in both forms, positive and negative eugenics do not differ merely in degree, but in kind. Each kind rests on a medically and morally distinct rationale. In negative eugenics, the aim is to prevent, control, or cure disease in people and thereby raise them to or maintain them at an absolute baseline of normal physical and mental functioning. In positive eugenics, the aim is to raise the level of functioning in people who are not diseased above the baseline. There may be relative differences in degree of disease or health below or above the baseline. But the baseline itself is an absolute measure that distinguishes between disease and health, as well as the aims of treatment and enhancement. It does not admit of degrees. If this is correct, and they are conceptually distinct, then negative eugenics is not assimilable to positive eugenics. One does not cause or inevitably lead to the other, and therefore we should reject any appeals to the slippery-slope argument to provide reasons for banning human cloning.

The alleged slide from negative to positive eugenic cloning need not occur. We can formulate public policy and enact legislation that upholds a

clear distinction between promoting health to a level of normal functioning, on the one hand, and enhancing people's capacities above this level, on
the other. These are distinct aims generated by completely different sets of
reasons corresponding to beneficence and perfectionism. A similar policy
works quite well in Britain, where there is a generally respected regulatory
authority that oversees IVF, human embryo, and stem cell research. Provided that the rationale for cloning is clearly spelled out at the policy level
in terms of health promotion through disease prevention, there is no reason to believe that cloning would inevitably lead to an enhancement program that would make some people unfairly better off than others, or that
it might adversely affect our humanity.

Conclusion

The possibility of cloning human cells, tissues, organs, and perhaps even
full-fledged human beings offers great opportunities for preventing, controlling, and curing genetically caused diseases. Although it still may be a
remote biological possibility, the moral implications of cloning human organisms that would develop into persons make us well-advised to proceed
cautiously on this front. For if we clone humans solely for the purpose of
making genetic replicas of ourselves, to replace a dying person, or as a
source of stem cells or bone marrow to save another person's life, then the
practice would violate a fundamental moral principle. The clone would be
used solely as a means and not also as an end, in which case cloning would
be morally objectionable. It is in this respect that cloning would undermine one's autonomy, dignity, and humanity. Nevertheless, cloning does
not necessarily imply such a consequence, since clones may very well be
treated as ends and not just as means. So there is nothing inherently
morally objectionable about cloning itself. The moral issues pertain to the
ways in which cloning is used.

Cloning should be medically and morally permissible when it is motivated by and used for therapeutic goals. Yet, to the extent that cells can be
genetically reprogrammed to produce genetically identical but healthy
copies of diseased tissues and organs, the cases in which there would be
compelling medical or moral reasons to clone human beings would be
rare. Indeed, this practice would have the virtue of sidestepping the most
troublesome moral questions generated by cloning. Cloning cells, embryos, or organs to replace diseased body parts would be either morally
permissible or morally neutral, since these parts have no moral status of

their own. Nor would cloning threaten our personhood or personal identity, because persons are constituted by but are not identical to their body parts. Genetic identity is not personal identity; unlike cells, selves cannot be cloned. Personhood and personal identity are not simply functions of tissues, organs, or genes. Provided that it is consistent with the idea of negative eugenics, of promoting health by preventing and treating disease, cloning can be a morally justifiable form of genetic technology that accords with our deepest metaphysical and moral convictions about the nature of persons and the value of our lives.

Assuming that the Kantian injunction to treat clones not solely as means but also as ends is upheld, there are only two morally objectionable aspects of cloning. The first is that the procedure might entail DNA damage to cell nuclei, which in turn could lead to premature aging and disease in people who have been cloned. The second is that, as an asexual form of reproduction practiced on a broad scale, it might entail long-term adverse effects on genetic diversity and the survival capacity of the human species. Although this scenario may not be likely, many people might avail themselves of it to minimize their children's susceptibility to disease. It could have adverse intergenerational effects by increasing the number of genetic mutations within families. Among other things, it raises the question of weighing the potential benefits of cloning and other forms of genetic manipulation and transfer between people who exist now and in the near future against the potential harms to people who will exist in the distant future. This moral weighing is necessary because manipulating genes could adversely affect genetic diversity, the result of which could be a higher incidence of deleterious mutations causing a higher incidence of disease and death earlier in people's lives. Another issue that raises similars questions is the possibility of genetically manipulating the mechanisms of aging to extend the human life span. It is important now to examine the evolutionary biological factors involved in altering genes for this purpose, as well as the metaphysical and moral consequences of it for both present and future people.

5

EXTENDING THE
HUMAN LIFE SPAN

In the preceding four chapters, I have examined different types of genetic intervention and have explored their implications for the existence and lives of future people along biological, metaphysical, and moral dimensions. My aim in this chapter is to further elaborate these issues along these same dimensions, paying special attention to the possibility of controlling the genetic mechanisms of aging and extending the length of the human life span. The discussion will further elucidate the impact of genetic intervention on the balance between acute and chronic diseases, personal identity through time, and our obligations to future generations.

Over the last hundred years, the developed world has experienced a substantial decline in mortality and a corresponding increase in life expectancy. This is due largely to the epidemiological shift from the predominance of acute infectious diseases earlier in life to chronic degenerative diseases later in life.[1] Nevertheless, even if the most common causes of death—cancer, heart disease, and stroke—were eliminated, the increase in life expectancy would be no more than about fifteen years. Moreover, although the *average* length of life (life expectancy) in developed counties has steadily increased, the *maximum* duration of life (life span) has not. Centuries ago, few people lived 100 years, and today this limit remains fundamentally unchanged. There are more centenarians today, but they constitute a very small percentage of the human population.

One of the essential biological properties of humans is senescence, the process by which cells, tissues, and organs gradually deteriorate as we age. This biological aging entails the decreasing ability of molecular mechanisms to properly maintain cell function. Aging is an inescapable fact of our biological lives and has different explanations. As I noted in chapter 4 with respect to the Hayflick limit, one explanation is that cells follow a strict genetic program and divide only a certain number of times before dying. Another account, also noted in chapter 4, is that over time mutations accumulate in cells and adversely affect the mechanisms controlling cell maintenance and repair.[2] Yet another explanation is that unrestricted caloric intake affects glucose metabolism in such a way as to increase the production of free radicals, which in turn damage cells. There is no single account of aging because it is governed by many different genes and molecular processes. But all of these processes involve senescence. Except for a small number of individuals, senescence ensures that humans cannot survive much beyond 100 years.

Although we have not ascertained all of the biological factors in aging, it is worth considering the role of telomeres in this process in the light of recent genetic research. As described in chapter 4, telomeres are the ends of chromosomes consisting of repeating DNA sequences that control the number of cell divisions and become increasingly shorter the more times cells divide. The telomerase gene is repressed in most somatic cells. In two separate experiments, however, researchers have discovered two ways to extend telomerase expression in these cells. One group was able to insert a copy of an active form of the telomerase gene into certain cell types and extend their life span in culture by at least twenty divisions beyond the Hayflick limit.[3] A different group was able to reconstitute human embryonic stem (ES) cells, which are undifferentiated and have permanently youthful telomeres.[4] In theory, cultured ES cells could be genetically programmed and guided into targeted tissues and organs to rejuvenate them. This could delay the onset of or ameliorate such age-related diseases as arthritis and diabetes. Moreover, if insertion of telomerase-active genes could make cells immune to the molecular clock that triggers senescence, and the life of cells were extended indefinitely, then it might be possible to extend the human life span another 50 to 100 years. These are rough figures without any intrinsic value. The point is simply to use them as references for reflecting on the biological and moral issues generated by the idea of a life span possibly twice the length of the present norm.

The idea of genetically manipulating the mechanisms of aging raises the following questions. If it became possible to extend the life span of our cells and bodies, then what would the motivation be for doing so? Would we want to maintain adult vigor for a longer period of time and delay the onset of or ameliorate the physical conditions entailed by senescence? Would we, in addition, want to exploit the technology in order to live much longer than we presently do? If one answered the last question affirmatively, then one would do well to consider the implications of the prospect of a significantly longer life span for one's identity and the quality of one's life. Equally important, one would do well to consider the effects that extending our lives in the present generation might have on the quality of the lives of the people who will exist in the next and succeeding generations. We need to reflect on what our obligations to them might be and what these would mean for the life-extending genetic interventions we might decide to undertake in the present generation.

Some remarks on the current public debate on embryonic stem cells are in order. While ES cells could be derived from cloned embryos, presently they are mined from IVF embryos that have been discarded by fertility clinics. In August 2001, President Bush announced that he would allow federal funding for limited stem cell research. This would involve existing stem cell lines from embryos that already had been discarded. No new embryos would be produced to generate additional stem cell lines.

The main moral issue here is weighing the value of allowing an embryo to develop into a full-fledged human organism and person against creating (and then destroying) it for research that could lead to treatment of degenerative diseases in existing people. We cannot benefit an embryo by allowing it to develop into a person because we do not thereby make the embryo better off. Embryos do not become persons, at least not in the identity-determining sense of "become." To insist otherwise is to conflate persons and human organisms, which are distinct ontological types. But we may be able to benefit people suffering from degenerative diseases by using ES cells to treat these conditions. We can make existing people better off by alleviating their pain, disability, and suffering. We have a stronger moral obligation to existing people than to merely potential people (embryos). These reasons would seem to justify using human ES cells for medical research and possible treatment. Yet, many believe that embryos and persons suffering from disease have equal moral value. The moral issues could be sidestepped by using adult stem cells from the skin, for example, rather than from embryos. But unlike these other cell types, ES cells are totipotent and therefore more versatile. So the debate on this issue is likely to continue for some time.

The issues that I am addressing in this chapter may not appear to be as immediately pressing as the moral status of the spare or cloned embryos from which ES cells are derived. But the idea of extending the human life span through genetic means needs to be addressed. For the interests of distant future people are no less morally significant than the interests of those of us living in the present and near future.

Consistent with the preceding chapters, the biological framework within which I will address these metaphysical and moral questions will be one informed by evolutionary biology. Insofar as natural selection determines the genetic mechanisms that control the life cycle of cells, there may be a good evolutionary explanation for senescence and death at the end of a limited life span. Natural selection allows for genetic mutations that cause disease later in life but keeps these mutations to a minimum earlier in life. The asymmetry in the way these mutations are selected offers a net reproductive advantage to human organisms so that they can transmit their genes to offspring. Deleterious mutations in the post-reproductive period are not selected against because they do not affect reproductive fitness. Yet substantially extending our biological life span through the insertion of genetically modified telomerase into, or reconstitution of ES cells at, the germ line could alter the course of natural selection, increase the number of deleterious mutations in humans earlier in life, and result in harm to distant future humans by making them more susceptible to premature disease and mortality.

I will explore how genetic intervention involving telomerase-active ES cells could alter the ratio of biological benefits earlier in life to biological burdens later in the life of human organisms. If this ratio were reversed, then extending the human life span well beyond the rough present limit of 100 years could mean that people existing in the distant future would have shorter and more diseased lives owing to a higher incidence of genetic mutations earlier in life. These people could be harmed by the defeat of their interest in existing without preventable early-onset infectious diseases and the opportunity for a decent minimum level of well-being over the course of a reasonably long life. Admittedly, this is a possible scenario, and as such does not warrant the stronger claim that all research aimed at extending the human life span should be foreclosed. Indeed, it would be implausible to make such a claim. For this research could lead to treatments for many degenerative diseases. Rather, I am making the weaker claim that, if one accepts the principles of evolutionary biology, then the possibility that I am raising should at least give us pause before we try to develop and implement life-extending technology on a broad scale.

In addition to the moral reasons I have just adduced, there are prudential reasons against genetically manipulating the mechanisms of aging to substantially extend the human life span. I will argue that a substantial increase in longevity would entail undesirable consequences for personal identity and the rational grounds for prudential concern about our future selves. A longer biological life of a human organism would not necessarily imply a corresponding continuous and unified psychological life of one and the same person. The point is not that a virtually immortal life of 200 years would be bad for us as individuals, but that it would not be prudentially desirable, because of its effects on personal identity. Yet, if it would not be desirable, then there would be no good reason for each of us to have a substantially longer life than we presently do.

Cells and Senescence

It is instructive to place the process of cell senescence and its effects on tissues and organs within the broader framework of the human immune system. We become more susceptible to disease as we age because cells divide only a finite number of times and because the underlying mechanisms that control cell damage and repair no longer function as efficiently and effectively as they did earlier in our biological lives. Apoptosis, or programmed cell death, is one factor that maintains equilibrium in the immune system.[5] Whereas the gradual shortening of telomeres and the number of cell divisions is written into cells from the point of conception, apoptosis is initiated in response to DNA damage affecting critical genes that control cell growth and proliferation. Apoptosis is necessary to maintain the delicate balance in the immune system between cell growth and damage repair to DNA, which is crucial to the survival of the organism. Too little apoptosis may lead to cancer and autoimmune diseases like rheumatoid arthritis; too much may result in stroke damage and neurodegenerative diseases like Alzheimer's.[6]

Immunological homeostasis is maintained by proteins that sense DNA damage from ionizing radiation, oxidation, or chemotherapeutic drugs. These proteins affect the cell cycle in either of two ways: (1) stopping cell division so that DNA damage can be repaired by the appropriate mechanisms; or (2) deciding that the damage has gone too far and that the cell must die. The tumor suppressor gene p53 is critical to the first mechanism in that it suppresses cell proliferation. If this response is overridden by oncogenes causing somatic-cell mutations, then the death-inducing protein tumor necrosis factor (TNF) is activated and causes the cell to die before it develops to the point of constituting a neoplastic risk. Ordinarily, when there is damage to DNA, the deregulated expression of dominant oncogenes that induce cell proliferation, such as Myc, E1-A, and E2F, is offset by the activation of p53 in response to DNA insult, as well as by key regulators of the cell cycle like the retinoblastoma protein (pRb). When these regulating factors are overridden by oncogenes, apoptosis (through activation of TNF) puts a stop to further deleterious proliferation. If apoptosis is effective, then the oncoproteins I have mentioned may indirectly act as tumor suppressors even when p53 has been activated. These features of the immune response constitute one way of ensuring that cells divide at a certain rate over a certain period of time.

It is important to emphasize the connection between cell senescence and apoptosis in the immune response. Inserting ES cells or manipulating genes to extend the period of telomerase expression in somatic cells could retard senescence and the onset of age-related diseases. It also could rejuvenate damaged tissues. At the same time, however, there is the risk that, by neutralizing or eliminating one of the controls on normal cell senescence, the altered cells could become cancerous. Paradoxically, a type of genetic intervention intended to extend the lives of cells and thereby extend adult vigor and the length of the human life span ultimately could turn out to shorten it. Immortalizing cells is a double-edged sword. To complicate matters further, not all cell types are equally sensitive to DNA insult or the function of telomerase. So it is unclear which cells would be more likely to respond to an active form of the telomerase gene in the desired way and which cells would be more likely to proliferate out of control and eventually kill us. In addition, if the investment that genes make in human reproduction comes at the expense of cell maintenance and repair, then, in the case of women who have borne children, it in unclear whether a telo merase gene in combination with ES cells could neutralize the effects of reproduction on cells and retard senescence in them.

Researchers have reported that, in experiments performed on mice, telomerase-immortalizing cells did not show any chromosomal changes. Moreover, even after two key growth suppressor genes, p53 and pRb, were activated, they did not find any cancer cells.[7] This suggests that crucial checkpoints on growth and proliferation are still intact in these cells. To be sure, one cannot make the simple inference that, because the experiments yielded these results in mice, they would yield the same results in humans. But even if we could have our cake and eat it too by retarding or even reversing the effects of senescence without simultaneously increasing the risk of cancer, there may be reasons against undertaking such a form of genetic intervention to lengthen the human life span well beyond the present norm.

Because the telomerase gene is activated in undifferentiated ES cells, these cells can divide indefinitely, provided that they remain embryonic. Only ES cells can remain undifferentiated and totipotent. As the embryo develops into a fetus and eventually into the body and brain of a full-fledged human organism, stem cells differentiate into the specialized cells necessary for the structure and continued function of the body's tissues and organs. The telomerase gene is expressed in undifferentiated cells, but it becomes repressed once differentiation occurs. Still, there is considerable variation among cells types. Brain and heart cells, for example, stop dividing once the organism is fully developed. In contrast, skin and blood cells continue dividing and can be rejuvenated by stem cells specifically targeted to these cell types. Once these types have differentiated, though, their cells have a finite life span, dividing only a limited number of times and eventually falling prey to senescence.

But a particular type of stem cell, present before embryonic development begins, can resist differentiation and the lifetime limit that this entails. These embryonic germ cells migrate to the developing ovary or testis, where they generate the egg or sperm for the next generation. By isolating these cells and making them immune to differentiation, researchers could ensure that the cells retained permanently youthful and active telomeres. If these ES cells with the active telomerase gene could be guided to carefully controlled specific pathways of differentiation, then it might be possible to "immortalize" the specialized cells of the blood, heart, and other organs.

There are weaker and stronger biological interpretations of "immortal." Neither of these interpretations should be taken literally, since neither implies that our lives can be eternal. Furthermore, these interpretations of immortality should be distinguished from the sense in which cancer cells are immortal, dividing endlessly out of control to bring about the demise of human organisms. According to the weaker sense of cell immortalization, genetically produced telomeres, or specially modified ES cells, could be introduced into somatic-cell types. This could moderately retard the process of senescence by rejuvenating cells, tissues, and organs and extending the period of adult vigor another ten to twenty years. I have referred to this as "regenerative medicine." According to the stronger sense of cell immortalization, undifferentiated telomerase-active ES cells, which are functionally equivalent to germ cells, could be programmed into the germ line at an early stage of embryonic development and guided into all cell types. Thus, even after differentiating and forming specific tissues and organs, all the body's cells could continue dividing in a youthful state for an indefinite period of time. This might make it possible for the human life span to be extended up to between 150 and 200 years. Of course, a longer life would be desirable provided that it did not include the physical, cognitive, and affective infirmities ordinarily entailed by aging, which the sort of genetic intervention in question presumably would prevent.

In many respects, there is no significant biological difference between extending the period of telomerase expression in somatic cells and many medical interventions currently in use. But each of these interventions allows at most a *moderate* extension of roughly ten to twenty years in the human life span, where the aim is not so much to increase the number of life years as to control diseases and maintain vigor for a longer period. A *substantial* extension of 50 to 100 years of the life span would be significantly different from current medical interventions because it could involve a life span twice the length of the present norm. Also, this extension would be motivated not just by the desire to remain healthy longer, but also by the deeper conviction that there is intrinsic value in living much longer than we presently do, given that being alive itself is intrinsically valuable.

Nevertheless, evolutionary biological reasons exist for not substantially extending our lives through genetic manipulation of the mechanisms of aging. And these biological reasons could influence moral reasons regarding how we should respect the interests of future people in not being harmed. Tinkering with genes at the germ line to extend life spans by reducing the incidence of genetic mutations later in life might gradually alter the course of natural selection. Such manipulation might increase the incidence of mutations earlier in life and hence the incidence of early-onset diseases in future people. Those who experience the initial effects of this type of intervention in the present and near future might benefit from it by having longer and healthier lives. But people who will exist in the distant future could be harmed by a "redressed" genetic balance involving a higher incidence of deleterious mutations earlier in life. Consequently, future people could be afflicted with a higher incidence of premature disease and mortality in their lives, which could be considerably shorter than our own.

Even if natural selection could correct itself and limit the number of mutations, this process operates at a glacial pace. Before a correction occurred, many people could be harmed by the defeat of their interest in having the same reasonably long life span that we presently enjoy. Insofar as we in the present generation have an obligation to prevent harm to people in distant future generations, and that extending our own lives through genetic manipulation of cells could harm these people in the respects I have described, we should carefully think through the implications of this technology before deciding whether, or to what extent, we should exploit it.

An Evolutionary Trade-Off

Aging itself is not a disease. But the diseases resulting from the gradual deterioration of the growth-and-repair mechanisms of cells are part of an age-related process. Generally, these will be chronic disorders of middle-to-late adulthood. Genes have different alleles, and alleles have different functions. The same genes with alleles that make human organisms susceptible to late-onset diseases also have alleles that protect us from early-onset diseases to ensure that we will survive long enough to reach reproductive age. This idea rests on the pleiotropic-gene hypothesis (which I will explain shortly). The disparate functions of alleles are part of one evolutionary purpose, which is ensuring that genes will be transmitted into the next generation. Since the most common way to do this is through sexual reproduction, natural selection determines that genes work to enhance the survival prospects of organisms until these organisms have reached reproductive age. Natural selection accomplishes this by limiting the number of deleterious mutations that would make organisms susceptible to diseases through reproductive age. Once the reproductive period during which genes would be transmitted into the next generation has passed, the survival and reproductive fitness of the organism have diminished importance. So natural selection will allow an increase in the number of mutations leading to a higher susceptibility to disease.

Natural selection maximizes the survival prospects of genes, not the health of human organisms or the well-being of persons. Nevertheless, because the health of human organisms is influenced by the ways in which genes control the cellular functions underlying physical and mental health, *which* genes (or alleles) are selected will influence the health of human organisms and the well-being of persons. Thus, although genes and the force of natural selection are impersonal, they are indirectly person-affecting.

The impact of natural selection on humans progressively weakens with increasing age, which suggests that genes providing biological benefits earlier in life may entail biological burdens later in life. The metabolic costs of maintaining the requisite mechanisms for human survival and reproductive success mean that mutations with harmful effects on human organisms in the later stages of their lives can accumulate within the gene pool. Given that survival and reproductive fitness are all that matter in evolutionary terms, there is no evolutionary reason for selection against mutations harmful to human organisms well beyond reproductive age. In the light of this, the question arises as to whether manipulating genes for the sake of longevity might adversely affect selection against mutations in reproductive and pre-reproductive years.

Human organisms have a limited lifetime biochemical budget, in the sense that their somatic cells divide only a certain number of times. The idea of a finite number of cell divisions is consistent with all of the explanations of aging that I cited at the outset of this chapter. Following the mandate of genes to be transmitted from one generation to the next, cells devote much of their energy in dividing to sexual reproduction. Insofar as a finite number of cell divisions implies that there is a limit to this energy, it seems to follow that cells will have less energy to devote to their maintenance and repair in the post-reproductive period. Even if the life cycle of cells could be extended through genetic manipulation, it is worth emphasizing that there is no evolutionary reason for natural selection to minimize mutations later in life. Manipulating genes to do this could alter the force of natural selection, increase the incidence of mutations earlier in life, and in turn affect the health and even survival of humans in reproductive and pre-reproductive years. This could result because we would be changing what the human body is naturally designed to do.

There may be an evolutionary trade-off between reproductive success and longevity, with resources invested to ensure reproductive success coming at the expense of a limited life span. When cells devote more of their energy to sexual reproduction, they have less energy to devote to maintenance and repair in the post-reproductive period. Conversely, if cells devote more of their energy to maintenance and repair for the sake of longevity, then less energy can be devoted to sexual reproduction. Since reproductive success ensures that genes will be transmitted into the next generation, it seems that the long life of genes entails a limited life span for humans.

Reproductive fitness is antithetical to premature disease and mortality in pre-reproductive and reproductive years, since diseases during these periods make it less likely that a human organism will survive up to the time when it can reproduce. The diseases in question are infectious in etiology, resulting from invading bacterial, viral, and other pathogens. These include such maladies as tuberculosis and HIV-AIDS. Because these conditions are caused by pathogens and are potentially lethal, I do not include the inability to reproduce in the same class of diseases, if indeed this could be considered a disease at all.

In the remote past, genes did not have to invest much in cell maintenance and repair because human organisms did not live much beyond reproductive age. All the biochemical energy went into promoting reproductive fitness, and cell mechanisms were part of the immune system's purpose to protect humans from pathogens and guarantee their survival so that they could reproduce. A recent study has shown that female longevity is negatively correlated with number of progeny and positively correlated with age at first childbirth. As the case of a childless woman (Jeanne Calment) in France who died recently at age 122 illustrates, a woman is more likely to live longer the fewer (if any) children she bears at a later age.[8] This supports the idea of an evolutionary trade-off between reproductive success and longevity.

Suppose that somehow we could manipulate genes so that they could enhance reproductive fitness *and* extend the human life span. We would do this by limiting the incidence of mutations in both earlier and later stages of life. If this were possible, then a longer life for present and near future people would not be incompatible with the survival and reproductive success of distant future people. Presumably, everyone would benefit and no one would be harmed. But these conclusions would not necessarily follow.

An extension of the life span would result in decreased adaptability of the human population because of increased competition for scarce resources, such as food, between older people, who already would have reproduced, and younger people. Perhaps younger people could adapt to this scenario by becoming more efficient in their caloric intake and metabolism during their reproductive years. But such adaptation would be difficult to achieve in a world where people's lives were significantly longer and where competition for resources like food could be quite fierce. Restrictions on procreation could alleviate the problem. Yet such a policy might be unfair to the young, who would not have the same reproductive choices that the old already have had. It is at least intuitively plausible that, in a future overpopulated world with substantially extended human lives, scarce resources could adversely affect the survival and reproductive prospects of the young. This could harm them by thwarting their interest in being healthy enough so that they could survive and procreate. The biological and moral implications of genetic manipulation of the mechanisms of aging cannot be separated from those of overpopulation, given an extended human life span.

According to the pleiotropic-gene hypothesis, which was introduced by George Williams, most genes involve both biological benefits and costs to human organisms.[9] This hypothesis is the foundation of evolutionary medicine, which understands the body and its responses to pathogens as a set of compromises that have evolved over time. In chapter 1, I cited Nesse and Williams's claim that evolutionary biology is the foundation for all biology, and that biology is the foundation for all medicine.[10] On this view, many genes have alleles that cause or predispose human organisms to one disease at the same time that they protect them against a different disease. The most well-known example of this is the allele for the autosomal recessive disease sickle-cell anemia. As already noted, one copy of the sickle-cell allele will protect against malaria, though two copies of the allele will cause SCA. This is an example of a gene that has been selected to confer a survival advantage on human organisms in an environment with a life-threatening pathogen, specifically the *Plasmodium* protozoan that causes malaria. Another example that seems to confirm the pleiotropic-gene hypothesis is the relation between cystic fibrosis and typhoid fever. Studying the relation between *Salmonella typhi*, the bacterium that causes typhoid fever, and the protein that, when inherited in the form of two copies of a mutant allele, causes CF, researchers have shown that *S. typhi* uses the healthy version of the CFTR protein to enter the gut cells it infects.[11]

Notably, like the relation between SCA and malaria, the relation between CF and *S. typhi* involves a trade-off between a genetically caused condition and a condition caused by a pathogen. In each case, natural selection allows the expression of an allele associated with the more chronic genetic condition because it protects against the more acutely life-threatening pathogen predominant in a given environment. This yields a net reproductive advantage to human organisms by allowing them to remain healthy and survive in different environments until they reach reproductive age. Strictly speaking, it is misleading to label genes as "healthy" or "unhealthy." While mutant alleles of genes can cause or make us susceptible to disease by adversely affecting the proteins regulating cell functions, the compromise between health and disease in human organisms is at bottom a reflection of genes working to be transmitted into the next generation.

Nesse and Williams claim that senescence is the most telling example of pleiotropy.[12] Genes that causally contribute to aging and eventually death later in life could be selected for if they gave an advantage to youth, when the force of natural selection is stronger. To the extent that the pleiotropic-gene hypothesis reflects a biological bias toward the earlier part of our lives over the later part, it is consistent with the explanation of aging in terms of mutations accumulating in humans during the post-reproductive period. Significantly, the converse hypothesis would be that genes "selected" to confer benefits later in life would entail more costly mutations earlier in life. The manipulation of genes for biological benefits in later years could, over many generations, result in a higher incidence of deleterious mutations in earlier years. The upshot of increasing longevity for ourselves in the present generation could be an increased incidence of acute early-onset diseases and premature death for people in the distant future. This would be due to the shift in the course of natural selection through manipulation of the genetic mechanisms of aging at the germ line.

Just because the scenario I have sketched pertains to the remote future does not diminish its moral importance. For, as Parfit has argued, the moral significance of an event is not determined by its timing. The fact that an event affecting people occurs in the remote future does not mean that it matters less morally, since the claims of all people not to be harmed have equal moral weight, regardless of the fact that they exist and make these claims at different times.[13] Parfit does say that "it may often be morally permissible to be less concerned about the more remote effects of our social policy. But this would never be *because* these effects are more remote. Rather, it would be because they are less likely to occur."[14] Presumably, then, the mere possibility of an increase in the number of mutations in the human gene pool should not occasion much moral concern.

Yet it is not so fanciful to imagine that many people would want to extend their lives if they had the means to do so. And if the mechanisms of aging were genetically manipulated in many people over many generations and altered the course of natural selection, then there could be a gradual increase in the number of deleterious mutations affecting humans earlier in their lives. The prospect of harm to future humans would become increasingly probable over time. This increasing probability, combined with the magnitude of the harm that the mutations entailed, provides grounds for carefully thinking through the moral implications of this type of genetic intervention. It is morally significant that people might be harmed in the distant future by having prematurely diseased and abbreviated lives.

As Nesse and Williams further point out, strong immune defenses involving macrophages, T lymphocytes, and the suppressor (p53) and necrosis (TNF) proteins protect us from infection by virulent microbes.[15] However, the long-term consequence of this protective immune response is low-level tissue damage that can lead to chronic diseases later on, such as arthritis, diabetes, and atherosclerosis. This claim is supported by a recent study concluding that atherosclerosis is an inflammatory disease that develops gradually through continued immune responses to viral, bacterial, or other pathogens.[16] Activation of monocyte-derived macrophages and T lymphocytes as part of the immune response can cause inflammatory arterial lesions in infants and children, which can signal the start of the process of atherogenesis. Similarly, in rheumatoid arthritis macrophages and lymphocytes predominate in the synovium, leading to the erosion of cartilage and bone, which is then replaced by fibrous tissue. A constant inflammatory response designed to protect the organism early in life can, over time, become an injurious response at molecular and cellular levels manifesting in diseases of the body's tissues and organs. In addition, *Helicobacter pylori*, the bacterium that plays a role in producing duodenal ulcers, may play a role in some stomach cancers. Such conditions are examples of an association between chronic disease and infectious agents

These sequelae may be unavoidable consequences of an immune system designed to protect human organisms from infection early in life to ensure their survival and reproductive fitness. This is an expression of the hypothesis that genes controlling cell function and immune response are pleiotropic. If our bodies in general and immune systems in particular are a delicate set of trade-offs favoring the earlier over the later stages of life, then it is possible that tinkering at the germ line with the genetic mechanisms of aging to extend the human life span for people in the present and near future could make distant future people more vulnerable to infectious diseases earlier in life.

One might claim that antibiotics and vaccines could prevent this scenario from occurring. But if there were more genetic mutations in the cells of human organisms, then it is likely that the cellular response of the immune system and its ability to neutralize infectious agents would be weakened. Antibiotics and vaccines are designed to stimulate or strenghten immune response to invading microbes. If the function of marrow-derived B cells and thymus-derived T (cytotoxic CD8 and helper CD4) cells has been compromised by genetic mutations, however, then one can question just how effective these interventions would be. For these cells are crucial in determining whether a human organism will have adequate immunological strength to fend off disease-causing pathogens. The increased susceptibility to infection of individuals whose immune systems are impaired

anisms of aging would be
essentially a form of en-
ventive medicine for oth-
quate physical and mental
priority to the claims of
e who would want to ex-
have a weak claim to us-
aims concerning enhance-
therapy and prevention.
ation of medical resources
nic stage of development,
ations leading to early- or
ond a certain point. Prior-
ation because it would be
afflicting people over the
r expensive treatments for
d be fair insofar as it gave
ong and relatively disease-
long and healthy lives ex-

e principle of diminishing
ditional unit of value will
ood than to someone who
han others because of dis-
ls by preventing disease in
of others who will live out
e former will have more
onsider their lives as com-
years for the healthy, or a
have no value for them.
achieve a normal life span
of living even longer than

n life span would bear on
e way it would impact on
derly citizens is increasing
or instance, the number of
to increase from the cur-
This will result in a larger

from the use of immunosuppressive drugs to reduce inflammation associated with autoimmune diseases, as well as individuals with HIV-AIDS, are
cases in point. Infectious agents evolve so fast that our immune systems are
always a step behind in responding to them. There is evidence of this in
the increasing number of antibiotic-resistant strains of bacteria, such as
methicillin-resistant *Staphylococcus aureus* (MRSA) and vancomycin-resis-
tant *Enterococcus* (VRE).

What I have described would mark a perverse twist in the current disease profile in modern medicine, with early-onset acute infectious diseases
becoming more prevalent than late-onset chronic diseases, and with people generally having increasingly shorter lives. There might even be a gradual decrease in the number of people who exist, given that fewer people
would reach reproductive age. Such a reversal would take a considerable
amount of biological time to evolve. Still, the point is not *when* it might
occur, but *that* it might occur. Present and near future generations of people might benefit from delaying senescence through genetic intervention.
But people in the distant future might be harmed by the defeat of their interest in having the same reasonably long lives without premature disease
that we presently have, as well as the same opportunities for projects and
achievements that give meaning to our lives.

In the light of these considerations and the possibility of immortalizing
human somatic cells through genetic manipulation of ES cells, telomeres, or
other molecular mechanisms of aging, it is instructive to cite Nesse and
Williams's words of caution. They say that "over the next decade, research
will surely identify specific genes that accelerate senescence, and researchers
will soon thereafter gain the means to interfere with their actions or even
change them. Before we tinker, however, we should determine whether
these actions have benefits early in life."[17] If the pleiotropic-gene hypothesis
is plausible, and if genetically manipulating the mechanisms of aging could
reverse the ratio of benefits to burdens between earlier and later stages of
life over many generations, then there are biological and moral reasons to
carefully assess how the lives of distant future people could be affected by
this intervention before endorsing the widespread use of it.

Intergenerational Equality and Fairness

The calculus of benefits and harms is not between different stages of one
individual's life, or the lives of a few individuals, but instead between the
probable benefits to some generations of humans and the probable harms

to other generations. Still, some might object to the co[n]
drawn. They might argue that the effects I have spelled [out]
concern would be so remote in the future and so unli[kely]
would be no reasons for prohibiting an increase in longev[ity]
netic manipulation. To support their claim, they might a[ppeal]
and invoke the Probabilistic Discount Rate, which says t[hat]
moral importance of future events is lower the less likely [to oc-]
cur.[18] On this view, a potential benefit to people in the pre[sent or near fu-]
ture would morally outweigh a potential harm to people i[n the remote fu-]
ture. But if more people availed themselves of this techno[logy, then over]
time there would be a greater probability of an increase [in longevity in]
the human gene pool. This probability, multiplied by the [amount of]
harm this would entail, is enough to take seriously the int[erests of]
future people.

The claims of some people not to be harmed have mor[e weight]
than the claims of other people to receive additional benefit[s. On the one]
hand, the additional benefit would be living beyond a no[rmal]
life span of eighty-five years or so. Since this would invo[lve]
longevity beyond the norm for the human species, it would [be odd to]
argue that not living beyond this limit would constitute har[m. Yet]
people living in the remote future could be harmed to the e[xtent that they]
would experience a higher incidence of disease earlier in life[. They]
would not be able to complete what we presently consider [a normal hu-]
man life span. For them, the harm would consist in the def[eat of their in-]
terest in not having to experience disease and living for a re[asonable num-]
ber of years with opportunities for a decent minimum level [of well-being.]

Present and future people are equal insofar as they do or w[ill have simi-]
lar claims not to be harmed by premature disease and limite[d op-]
portunities. Recall that fairness consists in meeting people's [needs in pro-]
portion to their strength, and that a fair distribution of a go[od]
gives priority to the strongest claims. Needs involve stronge[r claims than]
preferences, and the claims of some people to avoid harm hav[e greater]
weight than the preferences of other people for additional [benefits. Ac-]
cordingly, the claims of future people to avoid disease early [in life and to be]
able to have a life span as long as that of present people have [greater]
weight than the claims of the latter to have a life span longer [than the pre-]
sent norm. It would be unfair to ignore or give less weight to [the claims of]
future people to avoid harm and give more weight to the p[references of]
present people for additional benefits. Insofar as genetic ma[nipulation]

Moreover, genetic manipulation of the mec[hanism is]
expensive. Weighing the higher cost of what i[s genetic en-]
hancement for some against the lower cost of p[roviding oth-]
ers so that they can reach the critical level of ad[equate]
functioning further supports the claim of givin[g priority to]
those who have or would have worse lives. Tho[se who ex-]
tend their lives beyond a normal life span woul[d be us-]
ing scarce medical resources for this purpose. C[laims for enhance-]
ment always are weaker than claims concerni[ng treatment.]
Even if we restrict the question of a fair distri[bution of benefits]
and costs to genetic intervention at the embry[onic stage,]
there would be stronger reasons to correct mu[tations causing]
late-onset diseases than to increase longevity be[cause prior-]
ity should be given to the first type of interve[ntion as]
more cost-effective in preventing diseases from[running the]
course of their lives, thus minimizing the need [to treat]
chronic conditions. In addition, the policy wo[uld give]
some people the opportunity to have a life as [long and disease-]
free as others who want their already sufficient[ly long lives ex-]
tended further.

These claims can be reinforced by appeal to [the diminishing]
marginal value of life years.[20] Generally, an a[dditional unit will]
matter less to someone who has more of some g[ood than to one who]
has less of it. If some people have less life years [due to dis-]
ease, then we should give priority to their nee[ds and benefit]
them, as against extending the number of years [of those who have]
a normal biological life span. More years to [the former have more]
value than more years to the latter when we [compare com-]
plete units. This is not to say that additional li[fe years, or a]
delay in the onset of age-related diseases, wou[ld have no value.]
But the point is that, in general, being able to [avoid disease]
has more value for one person than the value [of avoiding]
this for another person.

Another problem is how extending the hum[an life span affects]
the problem of overpopulation, especially in t[erms of]
the quality of people's lives. The proportion of [older people is increasing]
in industrialized nations. In the United States, [the number of]
persons over sixty-five years of age is projecte[d to grow from the cur-]
rent 30 million to 58 million by the year 203[0]

number of nonworking older citizens being supported by a smaller number of working younger citizens (given the declining birthrate). The latter will have to bear the burden of paying higher payroll taxes to support extended reliance on such programs as Social Security and Medicare. More accurately, the problem is not overpopulation as such, but rather the disproportion between the numbers of people in different age groups.[21]

There would be additional disturbing consequences if the life span of existing people were substantially extended. More people would be living at the same time. An increase in longevity for people who already exist, without a significant decrease in the number of people coming into existence during the same period, would further strain the availability of already scarce resources like food, water, clean air, and health care. Without an increase in the availability of these and other resources, the quality of life for all people who exist at the same time could be lowered significantly. Some might maintain that population size has not diminished the availability of food, owing mainly to the fact that science has been able to sustain the carrying capacity of the planet. Agricultural technology has been able to keep pace with demand in developed and most developing counties, where population has been steadily increasing. Fossil fuels are largely responsible for our ability to develop fertilizers to cultivate land, sow and harvest crops, and deliver food to consumers. Yet fossil fuels also constitute a large portion of the carbon dioxide emitted into the atmosphere, the level of which is expected to double by the year 2040 and likely become much worse in the further future. So, the technology that we have used to maintain adequate food production for the population already may have caused, and may continue to cause, adverse changes in climate. This is turn could have an adverse impact on agriculture and result in diminished food production. Increasing people's life span could only exacerbate this problem.[22]

Limiting procreation would be one way of addressing the problems associated with overpopulation. But it would force people to make a difficult choice. In Kavka's words, "it is therefore possible to *imagine* a future world in which population pressure and advanced genetic technology combine, such that society, to protect itself, will offer individuals a harsh choice between (1) extending their own life span, at the price of surrendering the right to reproduce, or (2) retaining the right to reproduce without having an extended life span."[23] Still, the paradox here is that, while this does seem to be the inevitable choice given competition for scarce resources, the elderly with increased longevity who might not have worked for many years would require a substantial number of younger individuals to support them in old age. The sorts of goods for which the elderly would be competing within their own cohort (e.g., health care, pensions) would be different from the goods for which younger adults would be competing within theirs (e.g., education, jobs). Partly because of this difference, lowering the population of people who existed in the same 150-year period by reducing the number of people who came into existence would have a negative impact on the well-being of the very old if there were not an adequate number of working young to support them. Thus, although an increase in human longevity would result in more competition for scarce resources and would likely lower the quality of life for all people who exist at the same time, the quality of life of the very old could be the most adversely affected.

Depending on the resources in question, in a world of many people with an extended life span, the number of additional people who come into existence could be both beneficial (paying to support Social Security and Medicare) and burdensome (competition for water, food, and health care) to those who already have lived many years. On the other hand, longer lives for those who already exist would be unduly burdensome to those who come into existence, who would have to support them. Moreover, the perception of unfairness among the young could result in subtle forms of discrimination against the old. Also, if our health care system adopts the view that all the stages of one's life have equal weight, rather than later stages having more weight than earlier ones, then those who have much longer lives may suffer when senescence finally catches up with them.[24] People's physical, cognitive, and emotional needs are greater in the last than the middle stage of their lives, and a life span approach to meeting needs may result in the old not receiving adequate medical treatment.[25]

We might decide to give priority in the allocation of health care to medical needs of people in the last stage of an extended life span. But this would mean that the medical needs of other people in the early and middle stages of their lives would not adequately be met. Rationing health care that gives weight to the needs of all people requires limiting health-care spending on the very old. Daniel Callahan articulates this problematic feature of extended lives:

> However far one goes here, there will always come a point where sickness will appear once again, for the diseases and disabilities cured earlier in life—even earlier in old age—are bound to be replaced by later sicknesses and disabilities, which must also in turn be coped with. Since it is impossible to get out of this world alive, and ordinarily impossible to get out of this world without being

sick prior to death, we can be certain that there is no way of avoiding in the end some form of rationing for the elderly.[26]

In this and other respects, extending the human life span simply may be putting off the inevitable and not making us collectively any better off than we are within a normal life span. It may even make us collectively worse off than we otherwise would have been.

Recall Kavka's point about the compromise between increased longevity and the right to reproduce. Perhaps the deepest irony here is that the choice between an extended life span and the right to reproduce may very well be made not *by* people but *for* people. Given the evolutionary trade-off between longevity and reproduction discussed in the last two sections, over the course of many generations a longer life may mean surrendering the *ability* to reproduce, not just the *right* to reproduce.

Taking stock, my main aim in this section has not been to generate Malthusian worries about overpopulation and what it might imply for the planet's carrying capacity and the human species' ability to sustain itself. Rather, I have made two points. First, the quality of life for people with extended life spans would be contingent upon a critical number of working adults. But this number could not be so high as to increase competition for scarce resources equally valued by all people and result in a decrease in the quality of life for both young and old living at the same time. It would be extremely difficult to specify what the optimum population would be, a problem made more difficult by the fact that certain goods, such as health care, have more value for older people than for younger ones, who have less need of them. Second, it would be unfair to the young, who would have to sacrifice a significant amount of income and other goods to support the needs of the old. The unfairness to the young would increase as the number of people coming into existence gradually decreased over successive generations, consistent with the inverse relation between longevity and reproduction. These points, in addition to the discussion of evolutionary biology in the earlier sections of this chapter, provide grounds for not extending the human life span through genetic manipulation of the mechanisms of aging.

Prudential Concern About the Future

Suppose that none of the harmful collective effects on future people that I have described would result from an increase in longevity. Science would be able to address population problems and there would not be an increase in the number of genetic mutations in the human gene pool. Future generations would not be made worse off than present ones. Still, there would be prudential reasons against the prospect of a much longer life for individuals. It would not be in individuals' best interests to have such a life because it would entail a disconnection of the mental states that form personal identity through time. Because the psychological connectedness in which personal identity consists is a matter of degree and holds only for so long, the distant future selves in an extended life span would be distinct from the present selves who decided to extend their lives. The interests of a distant future self would be distinct from those of my present self because that future self would not be me. Hence, a life substantially longer than 100 years would not be desirable. By "prudential concern" I mean the concern individual persons have about their *own* future. In discussing this, I will ignore such things as concern about one's distant descendants or the future condition of the environment.[27]

We can imagine that many individuals would choose to extend their lives if the genetic technology to do this became available. Those undergoing the procedure could have a prolonged period of vigor before the effects of senescence set in and probably a moderate extension of life beyond a normal span of about eighty-five years. They might also be able to live substantially beyond this point. Let us consider each of these scenarios in turn.

The motivation for the first scenario would not be to increase the number of life years as such, but rather the number of years of vigor and thus the quality of these years. Presumably, this would imply a corresponding decrease in or compression of the number of years between the onset of age-related diseases and death.[28] Ideally, one would want to delay the onset of disease until the completion of most or all of one's projects within a life plan. At this point, one still would be physically, cognitively, and emotionally able to enjoy one's achievements and die not too much later from an acute illness such as pneumonia. The motivation for the second scenario would be the conviction that being alive is intrinsically good and therefore that living longer is intrinsically better than living for a shorter period. On this view, there is no point beyond which additional years of life have diminishing marginal value. This implicitly assumes that in being alive a person is free from the pain and suffering occasioned by disease and disability and has the capacity and opportunity to undertake and complete projects within a reasonably long life span. The value of a longer life will be conditional upon the capacities of the person whose life it is. In order to determine this, however, we need to reiterate the biological and psychological senses of "life" and the criteria of personal identity spelled out in Chapter 1.

The first sense of "life" is biological and pertains to the persistence of a structurally and functionally integrated human organism. The second sense of "life" is psychological and pertains to the capacity for mental life and the connectedness and continuity between mental states over time. Persons are constituted by their organisms but are not identical to them, since they are essentially defined in terms of psychological rather than biological properties. Although the life span of a person overlaps the life span of an organism to a certain extent, the two entities are not coextensive. Human organisms have longer life spans than persons, beginning earlier and ending later than persons. So human organisms and persons are ontologically distinct kinds. Psychological connectedness and continuity form the criteria of personal identity. The first relation consists in the holding of particular direct links between mental states. These connections can be stronger, holding over shorter periods of time, or weaker, holding over longer periods. The second relation is the ancestral relation of psychological connectedness, consisting of overlapping chains of strong connectedness and extending over longer periods of time than what is involved in particular direct links between mental states. Unlike connectedness, continuity does not admit of degrees and is a transitive relation.

Strictly speaking, personal identity is based on the relation of continuity rather than connectedness. Parfit maintains that connectedness pertains to "what matters" but is not equivalent to identity.[29] Yet, in an intuitive sense, what matters to us and our conception of ourselves as the same individuals persisting through time cannot be separated. Accordingly, I will use the more intuitive sense of identity interchangeably with what matters to us and base both on the relation of connectedness. Specifically, I will be concerned with connectedness between forward-looking desires and intentions and backward-looking memories, unified from the standpoint of the conscious present.

The ontological distinction between human organisms and persons is crucial. For when we reflect on the desirability of a longer life, it is the continued conscious life of a person that we have in mind, not the mere continued biological functioning of a human organism. Few, if any, would find congenial the thought of an extended life span consisting in a persistent vegetative state, with our organisms sustained indefinitely by mechanical life support. Although the continued functioning of some parts of the body and some regions of the brain are necessary conditions of an extended biological life span, these regions are not sufficient for the continued conscious life of a person. Not all parts of the brain can generate and sustain mental life. In addition, the qualitative aspects of mental states, such as consciousness, and the contents of these states are not solely functions of the body and brain but also of the social and physical environments in which we exist. Nor can the body and brain alone fully explain how we integrate, or give narrative meaning to, our past- and future-oriented mental states from the conscious present. The continued conscious life of a person is a necessary condition of the rational desire for a longer life and of rational concern about one's future self.

Intuitively, the best psychological life, the one with the highest level of quality or well-being, will be one in which a person consistently has achievements for a good number of years, and where the level of these achievements declines just before death. As noted in chapter 2, most people have asymmetrical temporal attitudes, being more concerned about the future than about the past. We care more about the good experiences we might have in the future than about those we have had in the past, preferring the good to be in the future and the bad to be in the past. Given these asymmetrical attitudes toward the past and future, the ideal contour of a life would have the bulk of bad experiences in the early stage, most of our achievements in the middle to later stages, and few, if any, bad experiences associated with the pain and suffering of disease in the last, hopefully short, stage. If increased longevity entailed continued physical, cognitive, and emotional vigor and a retarding of senescence for a considerable period of time, then more life years would offer more opportunities for achievement, as well as opportunities to enjoy the fruits of these achievements through reflective memory. Such a life would have a very high level of value. On the other hand, the contour we want to avoid is one of an extended life span containing a disproportionately long last stage of disease and disability. In this case, a life with a prolonged final period of morbidity would make for a wretched last stage and might even give that life as a whole a net disvalue. This would be the case if, as a function of our temporal bias toward the future, we care more about avoiding bad experiences in the last stage of life than in any other stage.

Perhaps the best illustration of the state of affairs just described are the lives of Jonathan Swift's Struldbrugs, the immortals in *Gulliver's Travels*.[30] These individuals have endless lives that are worthless and intolerable, owing to the fact that more life years for them entail physical, cognitive, and affective deterioration, precisely the sort of misfortune that an extended human life span should avoid. Yet, even if genetic manipulation of the mechanisms of aging could ensure a life considerably longer than the present norm with one's capacities intact, the value of such a life could diminish beyond a certain point for the people who would live it.

For most of us, life plans, and projects undertaken within these plans, are made with the understanding that there is only so much that we can accomplish in life. There is a limit to the number of projects that can sustain our interest in persisting as persons into the future, as well as to maintain the same level of interest and enjoyment in experiencing projects indefinitely. This is due to the limit in our cognitive resources of knowledge and imagination. Moreover, this may explain both our temporal bias toward the *near* (as distinct from the *distant*) future and our idea of what the contour of a life should be. On this view, even if the extension of our lives indefinitely into the future did not entail the same infirm fate as Swift's Struldbrugs, it is doubtful that we could continuously generate a series of projects that could sustain our desire to go on living.[31] Beyond a certain point, continued life would be at best value-neutral, at worst intolerable, but most likely undesirable. In Thomas Hurka's words, "like a play, a life can have a denouement, but it should not take up two-thirds of its running time."[32]

One might object that, with a longer life span, we could devise different plans and projects that would have expanded temporal horizons. We could plan for a sequence of projects that would be internally related, each of which could constitute a lifetime project given our present maximum life span.[33] But our brains have evolved in such a way as to maintain equilibrium between anticipation of the future and memory of the past. Too much memory might compromise our ability to conceive of and undertake new projects, which depends on our ability to anticipate the future. Insofar as there are limits to what we can anticipate in the future and remember from the past, we could sustain our interest in any project for only a limited period of time. Because of these cognitive limits, and because our forward-looking desires and backward-looking memories would be only weakly connected over long periods of time, projects undertaken at much earlier and much later periods of life would be at most weakly connected for us. This should lead us to question just how internally related these projects could be. Later in this section, I will cite empirical research on memory to support this point.

In considering the possibility of a substantially extended life span, we need to ask how this would affect psychological connectedness between earlier and later mental states. Would a significantly long span of time between earlier and later mental states weaken the relations between them and make them so different that they effectively belonged to two distinct persons? Or would the relations between these states be strong enough to withstand the test of time and ensure that they were of one and the same person? When we say that we want to persist as persons into the future, it is not just qualitative similarity between present and future psychological states that we have in mind, but numerical identity. That is, the person who will exist in the future must be the very same individual who exists now. In anticipating or planning for my future, it is not enough that my present psychological states are similar to other psychological states at a later time. I assume that these states will be *mine* and that it will be *me* who exists at that time. Yet, I want such states as my earlier intentions and the acts in which the intentions are realized to be united by the stronger relation of connectedness, not the weaker relation of continuity. Parfit acknowledges that "connectedness matters more than continuity in theory and practice."[34] Psychological connectedness is necessary for what matters to us. Continuity does not admit of degrees and is transitive. Connectedness does admit of degrees, however, and over long periods of time diminishes and fails to meet the condition of transitivity necessary for personal identity in the strict sense.

If connectedness is a necessary ground of what matters to us in the future, and if connectedness diminishes over time, then in an extended life span it seems that we may have to settle for qualitative similarity rather than strict numerical identity as that ground. Furthermore, assuming that there is congruence between prudential concern about the future and the intuitive conception of personal identity through time, and that the psychological connections definitive of personal identity diminish over time, it follows that the reach of our prudential concern can extend only so far into the future. Because any mental states in the distant future would be so weakly connected to any present mental states, a substantial extension of one's biological and psychological lives would not be desirable from one's present point of view.

In such a scenario, the identity of one and the same structurally and functionally integrated human organism most likely would be preserved. Its biological properties would be relevantly similar between earlier and later times. But it is unlikely that the identity of one and the same person would be preserved, owing to diminished psychological connectedness over a long period. This is according to the intuitive conception of personal identity that I have adopted. Yet, even if one adopted a strict conception of personal identity based on psychological continuity, there would not be any reason for an earlier self to care about a much later self, since the connections necessary to ground what matters to us would have greatly diminished.

There would be a divergence of our biology from our psychology. But it is our psychology that provides rational grounds for concern about the future. The idea that our biological organisms might exist for up to 200 years, while the psychological relations in which personal identity through time consists became disconnected and discontinuous well before this point, would not be very comforting. A long enough biological life would entail distinct psychological lives of distinct persons. Trying to extend my psychological life as a person through genetic manipulation of the mechanisms of aging has the paradoxical consequence that, as my biological organism continued to exist indefinitely, I would cease to exist and a person distinct from me would begin to exist beyond a certain point. My earlier and later selves effectively would be the selves of different persons.

We might call this Methuselah's Paradox, after the biblical figure. Admittedly, the case of Methuselah is an exaggerated illustration of the problem of retaining or losing personal identity over an extended period of time. But David Lewis employs it instructively to show how the relations between earlier and later mental states can become disconnected and discontinuous for a person who lives too long:

> Consider Methuselah. At the age of 100 he still remembers his childhood. But new memories crowd out the old. At the age of 150 he has hardly any memories that go back before his twentieth year. At the age of 200 he has hardly any memories that go back before his seventieth year, and so on. When he dies at the age of 969, he has hardly any memories beyond his 839th year. As he grows older he grows wiser; his callow opinions and character at age 90 have vanished almost without a trace by age 220, but his opinions and character at age 220 also have vanished almost without a trace by age 350. He soon learns that it is futile to set goals for himself too far ahead. . . . For Methuselah the fading out of personal identity looms large as a fact of life. It is incumbent on us to make it literally true that he will be a different person after one and one-half centuries or so.[35]

This example shows how self-interested concern about the future is intimately related to the idea of persisting through time as the same person. The less psychological connectedness between the mental states of my present self and those of a remote future self, the less rational it will be for me to be concerned about that future self, which arguably would not be mine. Parfit offers some support for this point by saying that "connectedness is one of the two relations that give me reasons to be specially concerned about my own future. It can be rational to care less when one of the grounds for caring will hold to a lesser degree."[36] He further claims that we can "treat weakly connected parts of one life as, in some respects, or to some degree, like different lives."[37]

Suppose that I form a desire and intention to undertake a certain project and to complete it by a certain time in the future. When I do this in the present, I assume that the project I value now will be something that I still will value when I complete it. Moreover, I assume that, when I complete it, I will be able to remember my earlier desire and intention. For my concern about completing the project to be rational, there must be a unity or integration of the mental states of the desire and intention at the earlier time and the mental state of remembering my earlier desire and intention at the later time. What unifies these earlier and later mental states is my conscious awareness of myself as a being who persists through time as one and the same individual. But if the connection between my earlier desire and later memory weakens as the length of time between these mental states increases, to the point where I no longer identify with the desire or its content, then it seems to follow that my concern about the project, and the rational basis for it, can extend only so far into the future. Owing to the passage of time and its effects on the connections between mental states, there would not be enough psychological unity of earlier and later mental states for me to rationally care about a self and its projects in the distant future. One could justify this by explaining that the mental states of the future self would be the mental states of a different person.

It is true that if one is continuously involved in a very long-term project, then one's awareness of the project and its continuous status would have to be continuously updated. However, updating the project at different stages would not guarantee that one's mental states at the earliest and latest stages would be sufficiently related to sustain it over a very long period of time. The farther apart in time one's attitudes are, the weaker the connections between them will be, which could undermine the unity and continuity of the project, as well as one's interest in it.

This point can be supported by a biological explanation of how certain genes and structures of the brain control memory. Consistent with an evolutionary explanation of human biology, memory is designed in such a way as to enhance the survival capacity of human organisms so that they can reach reproductive age and transmit their genes to the next generation. As an essential part of a human organism, the brain will not allow the mind to become cluttered with memories that serve no purpose, especially if this interferes with other mental functions. Even if brain cells somehow could be rejuvenated through genetic manipulation in the same way as cells in other parts of the body, how the brain and mind form, store, and retrieve memories is determined ultimately by what events the brain deems useful for the organism as a whole. Short-term memory is essential to our synchronic sense of identity, to our conscious awareness of ourselves existing at a given time. Long-term memory is essential to our diachronic sense of identity, to our conscious awareness of ourselves persisting through time. The latter is especially germane to the congruence between personal identity and prudential concern about our future selves.

Memories the brain deems useful are archived via the hippocampus to a long-term storage site in the neocortex, the most recently evolved part of the brain. The action of a particular molecule—cyclic AMP response element binary protein (CREB)—maintains an optimal balance between learning and forgetting by modulating the formation-storage-retrieval process.[38] CREB is a transport factor responsible for switching on particular genes that manufacture proteins necessary for long-term memory formation. Crucially, CREB comes in two types: "activator" CREB and "blocker" CREB. The first activates the genes necessary for the formation of long-term memory, while the second interferes with the action of the activator type and thereby inhibits the formation of additional memories. Or, if memories have been formed already and are in storage, it removes them from mental circulation by interfering with the retrieval mechanism of recall. Too much blocker CREB results in forgetfulness and, in severe cases, dementia. Too much activator CREB results in an overproduction and oversupply of memory, the mind being cluttered with memories of events that serve no purpose. Because the hippocampus and neocortex would be so busy dealing with an overloaded storage system, it would be difficult to learn new things and anticipate events in the future. This would severely impair the planning and decision-making capacity necessary for human agency and personal identity.

The memory equilibrium achieved by an optimal balance between activator and blocker CREB is critical to the psychological unity of anticipation of the future and memory of the past, a unity critical to both personal identity through time and rational concern about the future. Yet the function of this protein suggests that the requisite unity of these mental states can hold only for a limited period of time. Anticipation cannot extend so far into the future that it undermines memory of the past. By the same token, there cannot be so much stored memory of past experience that it comes at the expense of our ability to anticipate and plan for the future. A break in this equilibrium would undermine the conditions of connectedness and continuity necessary for us to be aware of ourselves as the same individuals persisting through time. It also would undermine our ability to sustain long-term projects by undermining the unity of future- and past-oriented attitudes necessary to ground these projects.

One might argue that manipulating the genes that control the function of CREB could extend the period through which mental states can be connected. In that case, a longer biological life would not undermine the psychological relations required for the continued existence of persons. But it is worth repeating that our minds and brains are designed so that there is equilibrium between our memory of the past and our anticipation of the future, including our ability to learn new tasks. This equilibrium involves an optimal balance of mental states extending only so far into the past and only so far into the future. Altering this balance by the sort of genetic manipulation suggested could result in too little or too much memory, which could have disastrous consequences for our conception of ourselves as entities consciously aware of our existence at particular times as well as through time.

Memory suppressor genes and the inhibitory CREB they encode prevent "noise" that would interfere with our cognitive ability to concentrate and anticipate what we might experience in the future. The deleterious effects of excessive long-term memory are illustrated in Borges's character Funes, from the story "Funes, the Memorious."[39] Following an equestrian accident, Funes remembers every trivial detail of everything he experiences. Overloaded with memory, he becomes a prisoner of his own cluttered thoughts, unable to tolerate any additional experience. A greater capacity for memory that came at the expense of the ability to learn new things and have hopes and plans for the future would leave us stuck in the present and past as suspended selves.

If this account of memory is correct, then it affirms the claim that a person can persist through time as one and the same individual for only a limited period. Thus, from the idea that the life of a human organism can be extended indefinitely into the future, it does not follow that the life of the person associated with but not identical to the organism also can be extended indefinitely into the future. The genetic and biochemical features of our brains on which our mental life depends determine that the connections between mental states, like anticipation and memory, which ground personal identity hold only for so long and gradually weaken and fade away with the passage of time. Given the congruence between identity and prudential concern about the future, and given that the psychological connections between our present mental states and the future mental states of individuals existing 100 or more years from now would be weak, it would not be rational for us to care about these distant future selves.

If I decided to have the cells of my body and brain genetically manipulated so that I could have a much longer life, then the biological organism that existed in the future might be identical to the one of which I am presently constituted. But the person existing at the end of this process most likely would not be me because the requisite connections between past and future mental states for retaining identity would have weakened so much that they no longer would hold. The paradox here is that the motivation for the genetic manipulation would be to extend my psychological life as a person. Yet, in the end, what I achieved would be only an extension of the biological life of an organism that at most would be only weakly related to me. Extending the human life span substantially beyond the present norm would be undesirable because the person who underwent the genetic procedure for all practical purposes would have gone out of existence before his biological organism did. Although our mental life causally depends on the continued functioning of our biological life, it is the former that generates and sustains our desire to live longer.

The consequence I just described would not occur if the increase in longevity were moderate and did not extend too far beyond the limit of a normal biological life span. For, in that case, the temporal distance between present and future mental states would not be so great as to severely weaken the connections holding between them. The relations necessary to retain personal identity still would hold, assuming the continued normal functioning of the body and the relevant regions of the brain, as well as the genetic and biochemical controls on memory and other mental states. A *moderate* extension of the human life span could preserve the integrity of our mental life while sustaining biological vigor. The problem with trying to immortalize cells and thereby bring about a *substantial* extension of the life span is that the biological gain would entail a psychological loss, the loss of personal identity and the mental connections that ground what matters to us. Such a life would not be desirable for any person, and therefore there would not be any prudential reasons for substantially extending our lives.

Conclusion

If one's biological life were to be extended substantially into the future through genetic manipulation of ES cells, telomeres, or other mechanisms of aging, then one would want a corresponding extension of one's psychological life that ensured that one would persist as one and the same person through time. I have argued that, although the desired congruence between our biological and psychological lives could be ensured if the life span were extended moderately into the future, a substantial increase in longevity would severely weaken psychological connectedness over a vast temporal period. It would come at the expense of personal identity and rational grounds for concern about the future. Making the body's cells and tissues less susceptible to age-related diseases may result in longer and more vigorous lives for our organisms. But it does follow from this that the connections between our earlier and later mental states would be strong enough to ensure that the same person who underwent the genetic manipulation would still exist at the end of the process. A substantially longer life would not be bad for a person (in a sense, it *could not* be, if that person ceased to exist while his organism continued to exist); but it would not be desirable either. Therefore, there are no good prudential reasons for substantially extending the human life span.

Furthermore, even if our biological and psychological lives could converge in a desirable moderately extended life span, we would not simply want to retard senescence and the age-related diseases it entails. We would want to compress the period between the onset of these diseases and death. A life in which the vigorous functioning of cells, tissues, and organs was extended beyond the present rough limit would not be desirable if it were followed by a prolonged period of morbidity before death.

Although one could make a case for extending our individual biological and psychological lives moderately into the future, consideration of the collective consequences of doing so provides reasons for rejecting it. A world in which everyone's life span were extended beyond eighty-five years could increase competition for scarce resources and lower the quality of life for all. These collective effects in turn could make it worse for each person to have a longer life. It also could impose an unfair burden on the young, who would have to pay a considerable amount in payroll taxes and make other sacrifices to support the nonworking old, whose numbers could extend over several generations. Insofar as increasing longevity could be considered as a form of genetic enhancement, and insofar as the procedure would be costly, it would be more morally preferable to use resources to improve the lives of the large number of existing people whose quality of life falls below the critical level of a decent minimum. This would mean diverting resources from genetic technology to programs aimed at improving the health of poor, diseased people. The need-based claims of the worse off to have reasonably long lives have more moral weight than the preference-based claims of the better off to have longer lives.

In terms of evolutionary biology, the trade-off entailed by pleiotropic genes suggests that genetic manipulation of cells at the germ line to extend the number of times they divide runs the risk of increasing the number of deleterious genetic mutations in the human population. Altering genes to make us in the present generation less susceptible to chronic diseases later in life could make future people more susceptible to acute diseases earlier in life. There could be a gradual increase in the incidence of infectious diseases in younger people, which could result in a radical change in the curve of mortality and life expectancy over many generations. It could even mean that a significant number of people in distant future generations might not reach reproductive age, which would diminish the number of people who would exist.

The main issue here is a moral one. If the profile of disease were to shift as I have suggested, then people in the distant future would be harmed by the defeat of their interest in having a reasonably long and disease-free life. This harm would be the effect of our present genetic manipulation of the mechanisms of aging to extend our own lives. The fact that they do not presently exist but will exist in the distant future does not diminish the moral force of their claim not to be harmed. A mere difference in time does not matter morally.

Ultimately, it is this risk of harm to future people, in addition to the undesirable consequences for personal identity and prudential concern about our future selves, that provide reasons against genetically manipulating the mechanisms of aging to substantially extend the human life span. Accordingly, we would do well to relinquish our generocentrism and egocentrism and come to realize and accept the biological and psychological limits of our organisms and our selves. We should live by the wisdom of Psalm 90:12 and make the days we have in our actual limited lives count.[40]

NOTES

Introduction

1. These and other related issues are discussed in Mark Rothstein, ed., *Genetic Secrets: Protecting Privacy and Confidentiality in the Genetic Era* (New Haven, CT: Yale University Press, 1997). See also Diane Martindale, "A Pink Slip in the Genes," *Scientific American*, January 2001: 19–20.

2. See the essays in David Wasserman and Robert Wachbroit, eds., *Genetics and Criminal Behavior* (New York: Cambridge University Press, 2001).

3. Parfit, *Reasons and Persons* (Oxford: Clarendon Press, 1984), especially part 4; Heyd, *Genethics: Moral Issues in the Creation of People* (Berkeley: University of California Press, 1992).

4. Kitcher, *The Lives to Come: The New Genetics and Human Possibilities* (New York: Simon & Schuster, 1996); McMahan, *The Ethics of Killing* (Oxford: Oxford University Press, 2001).

5. *Why We Get Sick: The New Science of Darwinian Medicine* (New York: Times Books, 1994).

6. *From Chance to Choice: Genetics and Justice* (New York: Cambridge University Press, 2000).

7. *Ethics and the Limits of Philosophy* (Cambridge, MA: Harvard University Press, 1985), 6. The Latin term is *moralitas*; the Greek term is *ethos*.

8. *Sovereign Virtue: The Theory and Practice of Equality* (Cambridge, MA: Harvard University Press, 2000), 485, n. 1.

9. See, for example, F. M. Kamm, "Non-Consequentialism, the Person as an End-in-Itself, and the Significance of Status," *Philosophy & Public Affairs* 21 (Fall 1992): 354–389.

Chapter 1

1. See Victor McKusick, *Mendelian Inheritance in Man: Catalogue of Autosomal Dominant, Autosomal Recessive, and X-Linked Phenotypes*, Tenth Edition (Balti-

more: Johns Hopkins University Press, 1992). Introductions to genetics more pertinent to the philosophical issues I address include David Suzuki and Peter Knudtson, *Genethics: The Clash Between the New Genetics and Human Values* (Cambridge, MA: Harvard University Press, 1990); Daniel Kevles and Leroy Hood, eds., *The Code of Codes: Scientific and Social Issues in the Human Genome Project* (Cambridge, MA: Harvard University Press, 1992); Theresa Marteau and Martin Richards, eds., *The Troubled Helix: Social and Psychological Implications of the New Human Genetics* (Cambridge: Cambridge University Press, 1996); William Clark, *The New Healers: The Promise and Problems of Molecular Medicine in the Twenty-First Century* (New York: Oxford University Press, 1997); and *Human Genetics: Choice and Responsibility*, Medical Ethics Council of the British Medical Association (Oxford: Oxford University Press, 1998).

2. Ruth Hubbard and Elijah Wald, for example, discuss the range of effects of the external environment on the expression of genes in *Exploding the Gene Myth* (Boston: Beacon Press, 1997).

3. From an article whose title is the same as this citation, in *American Biology Teacher* 35 (1975): 125–129.

4. See U. S. Congress Office of Technology Assessment, *Mapping Our Genes: Implications for Health and Social Policy* (Washington, D.C.: National Academy Press, 1994). Maxwell Mehlman and Jeffrey Botkin offer a concise overview of the scientific, ethical, and policy aspects of the Genome Project in *Access to the Genome* (Washington, DC: Georgetown University Press, 1998), chap. 2. A more comprehensive discussion of the Project is given by Buchanan et al., *From Chance to Choice: Genetics and Justice.*

5. From R. Lipkin, "The Quest to Break the Human Genetic Code," *Insight*, December–January 1991: 46–48. Cited in Mehlman and Botkin, 14.

6. "Medical and Societal Consequences of the Human Genome Project," *New England Journal of Medicine* 341 (1999): 28–37. Also, Francis Collins and Victor McKusick, "Implications of the Human Genome Project for Medical Science," *Journal of the American Medical Association* 285 (February 7, 2001): 540–544, and the special issue of *Science* 291 (February 16, 2001), "The Human Genome." It should be noted that, although Venter and Collins announced that they had completed the mapping and sequencing of the human genome in June 2000, the results were not published until February 2001.

7. "Will Genetics Revolutionize Medicine?" *New England Journal of Medicine* 343 (July 13, 2000): 141–144, at 141. Parenthesis added.

8. Ibid., 141.

9. Alan Wolffe and Marjori Matzke, "Epigenetics: Regulation Through Repression," *Science* 286 (October 15, 1999): 481–486.

10. Ibid., 143, and "The Authors Reply," *New England Journal of Medicine* 343 (November 16, 2000): 1498. See also Richard Lewontin, *The Triple Helix: Gene, Organism, and Environment* (Cambridge, MA: Harvard University Press, 1999).

11. "Evolution and the Origins of Disease," *Scientific American*, November 1998: 86–93, at 86. This article is a precis of their book, *Why We Get Sick*, especially chapter 15. See also P. W. Ewald, *Evolution of Infectious Disease* (New York: Oxford University Press, 1994), and W. R. Trevathan et al., eds., *Evolutionary Medicine* (New York: Oxford University Press, 1999).

12. See, for example, Peter Delves and Ivan Roitt, "The Immune System," parts 1 and 2, *New England Journal of Medicine* 343 (July 6, 2000): 37–49, and 343 (July 13, 2000): 108–117; Ruslan Medzhitov and Charles Janeway, "Innate Immunity," *Ibid.*, 343 (August 3, 2000): 338–344; Roitt, *Essential Immunology*, Seventh Edition (Oxford: Blackwell Scientific Publications, 1991); and Marc Lappe, *The Tao of Immunology* (New York: Plenum Press, 1997).

13. Proponents of this view include Christopher Boorse, "'Health' as a Theoretical Concept," *Philosophy of Science* 44 (1977): 542–571, and Leon Kass, "The End of Medicine and the Pursuit of Health," in *Toward a More Natural Science: Biology and Human Affairs* (New York: Free Press, 1985): 157–186. A much broader (too broad, I believe) definition of "health" is given in the preamble to the Constitution of the World Health Organization (1986): "'Health' is a state of complete physical, mental, and social well-being and not merely the absence of disease or infirmity."

14. "The Genome Project, Individual Differences, and Just Health Care," in T. Murphy and M. Lappe, eds., *Justice and the Human Genome Project* (Berkeley: University of California Press, 1994): 110–132, at 122. Daniels first formulated this definition of health in *Just Health Care* (New York: Cambridge University Press, 1985).

15. James Sabin and Norman Daniels, "Determining 'Medical Necessity' in Mental Health Practice," *Hastings Center Report* 24 (1994): 5–13, at 10.

16. Preben Bo Mortensen et al., "Effects of Family History and Place and Season of Birth on the Risk of Schizophrenia," *New England Journal of Medicine* 340 (February 25, 1999): 603–608, and Nancy Andreasen, "Understanding the Causes of Schizophrenia," *Ibid.*: 645–647.

17. Boorse, "On the Distinction Between Disease and Illness," in M. Cohen et al., eds., *Medicine and Moral Philosophy* (Princeton, NJ: Princeton University Press, 1981): 3–22. Also, Eric Cassell, *The Nature of Suffering and the Goals of Medicine* (New York: Oxford University Press, 1991).

18. Kitcher discusses these two models of disease in *The Lives to Come*, chap. 9, "Delimiting Disease."

19. Ibid., 216. Emphasis added.

20. Hubbard and Wald, *Exploding the Gene Myth*, 85.

21. National Center for Health Statistics: 1979–92. Cited in *Scientific American*, October 1995, 32D.

22. *Why We Get Sick*, and "Evolution and the Origins of Disease."

23. Mel Greaves argues that cancers are the ends of long, unbroken evolutionary chains that extend back billions of years in *Cancer: The Evolutionary Legacy* (New York: Oxford University Press, 2000).

24. Clark, chap. 3, Melhman and Botkin, chap. 2, and Lee Silver, *Remaking Eden: How Genetic Engineering and Cloning Will Transform the American Family* (New York: Avon Books, 1998), chap. 1.

25. Antonio Damasio explains this role of the body in mentality in *Descartes' Error: Reason, Emotion, and the Human Brain* (New York: Grosset/Putnam, 1994), chap. 10, "The Body-Minded Brain," and in *The Feeling of What Happens: Body and Emotion in the Making of Consciousness* (New York: Harcourt Brace, 1999).

26. See Thomas Nagel, "What Is It Like to Be a Bat?" in *Mortal Questions* (New York: Cambridge University Press, 1979): 149–163; John Searle, *The Rediscovery of the Mind* (Cambridge, MA: MIT Press, 1992); Colin McGinn, *The Mysterious Flame: Conscious Minds in a Material World* (New York: Basic Books, 1999); and Todd Feinberg, *Altered Egos: How the Brain Creates the Self* (New York: Oxford University Press, 2001), chap. 9.

27. Parfit, *Reasons and Persons*, 202.

28. Ibid., 211 ff. Also, David Lewis, "Survival and Identity," in A. O. Rorty, ed., *The Identities of Persons* (Berkeley: University of California Press, 1976): 17–40, and Peter Unger, *Identity, Consciousness, and Value* (New York: Oxford University Press, 1990). In *The View from Nowhere* (New York: Oxford University Press, 1985), Nagel defines personal identity in terms of psychological continuity, though he traces this ultimately to the brain alone. "Where the brain goes, I go" (41). Parfit and Lewis maintain that psychological connectedness and continuity can occur in branching forms and appeal to split-brain and other thought-experiments to support their position. This underlies their claim that identity as such does not matter. Unger and Nagel hold that continuity does not allow for branching but requires a single unity of consciousness persisting through time. I will assume that psychological connectedness and continuity are nonbranching. Furthermore, I will assume that the particular

brain and body an individual has are essential for identity; not just any brain or body will do.

29. Prominent contemporary materialists include David Armstrong, *A Materialist Theory of Mind* (Cambridge: Cambridge University Press, 1969); Paul Churchland, *Matter and Consciousness* (Cambridge, MA: MIT Press, 1984); and Sydney Shoemaker, "Personal Identity: A Materialist's Account," in Shoemaker and Richard Swinburne, *Personal Identity* (Oxford: Blackwell, 1984). The most prominent contemporary dualist is Swinburne, "Personal Identity: The Dualist Theory," Ibid., and *Evolution of the Soul*, Revised Edition (Oxford: Clarendon Press, 1997).

30. Parfit, *Reasons and Persons*, 211.

31. Ibid., 213–214.

32. In "Human Beings," *Journal of Philosophy* 84 (February 1987): 59—83, Mark Johnston argues that "human organism" names a purely biological kind, while "human being" names a partly psychological kind (79). I disagree with this distinction because the mere *potential* for psychological life does not make something even partly a psychological kind. It must possess the *capacity* for psychological life, and only persons have this capacity. Michael Tooley, for one, criticizes the notion of "potential persons" in *Abortion and Infanticide* (Oxford: Clarendon Press, 1983), chap. 6. In *Causing Death and Saving Lives* (Harmondsworth: Penguin, 1977), Jonathan Glover maintains that a fetus is a potential person, and John Harris calls embryos "potential persons," or "pre-persons" in *Clones, Genes, and Immortality* (Oxford: Oxford University Press, 1998), 8. Yet Stephen Buckle persuasively argues for the distinction between the potential to *become* and the potential to *produce* in "Arguing from Potential," *Bioethics* 2 (1988): 227 ff. I follow Buckle in claiming that zygotes, embryos, and early-stage fetuses only have the potential to produce persons, which does not imply the logical relation of identity holding between these different stages of human development, as implied by the potential to become persons.

33. A similar position is defended by Lynn Rudder Baker in *Persons and Bodies: A Constitution View* (New York: Cambridge University Press, 2000). A sustained defense of the view that we are essentially organisms is given by Eric Olson in *The Human Animal: Personal Identity Without Psychology* (Oxford: Clarendon Press, 1997). Peter Van Inwagen argues in the same vein in *Material Beings* (Ithaca, NY: Cornell University Press, 1990).

34. David Wiggins, *Sameness and Substance* (Oxford: Blackwell, 1980); David Oderberg, "Coincidence Under a Sortal," *Philosophical Review* 105 (1996): 145–171; and "Modal Properties, Moral Status, and Identity," *Philosophy &*

Public Affairs 25 (Fall 1997): 259–298; and Jeff McMahan, *The Ethics of Killing*, chap. 1.

35. McMahan, *The Ethics of Killing*, chap. 1; Silver, *Remaking Eden*, 69; and Stephen White, *The Unity of the Self* (Cambridge, MA: MIT Press, 1989).

36. Buckle, Karen Dawson, and Peter Singer, "The Syngamy Debate: When Precisely Does a Human Life Begin?" in Singer et al., *Embryo Experimentation* (Cambridge: Cambridge University Press, 1990): 213–225. Also, Harris, *Clones, Genes, and Immortality*, 61–65; Michael Lockwood, "When Does a Life Begin?" in Lockwood, ed., *Moral Dilemmas in Modern Medicine* (Oxford: Oxford University Press, 1985): 9–31; and Norman Ford, *When Did I Begin?* (Cambridge: Cambridge University Press, 1988), 97 ff.

37. *Human Embryos: The Debate on Assisted Reproduction* (Oxford: Oxford University Press, 1989), 17.

38. Silver, *Remaking Eden*, 64.

39. See, for example, J. A. Needham, *A History of Embryology* (New York: Abelard-Schuman, 1959); C. Crobstein, "Cytodifferentiation and Its Controls," *Science* 143 (1964): 643–650; and Gerald Edelman, *Neural Darwinism* (New York: Basic Books, 1987).

40. Buckle, "Arguing from Potential."

41. "Lewis, Perry, and What Matters," in Rorty, 103, n.10.

42. See W. French Anderson, "The Best of Times, the Worst of Times," *Science* 288 (April 28, 2000): 627–629, R. M. Blaese et al., "T Lymphocyte-Directed Gene Therapy for ADA-SCID: Initial Trial Results After Four Years," *Science* 270 (1995): 475; and Rebecca Buckley et al., "Hematopoietic Stem-Cell Transplantation for the Treatment of Severe Combined Immunodeficiency," *New England Journal of Medicine* 341 (February 18, 1999): 508–513.

43. Marina Cavazzana-Calvo et al., "Gene Therapy of Human Combined Immunodeficiency (SCID)—X1 Disease," *Science* 288 (April 20, 2000): 669–672.

44. The most perspicuous account of harm is developed and defended by Joel Feinberg in *Harm to Others* (New York: Oxford University Press, 1984) and *Harm to Self* (New York: Oxford University Press, 1986).

45. Feinberg, Ibid.; Parfit, *Reasons and Persons,* chapt. 15, and John Broome, *Weighing Goods* (Oxford; Blackwell, 1991), chapt. 8.

46. See, for example, the discussion of the "principle of parental responsibility," by Bonnie Steinbock and Ron McClamrock, "When Is Birth Unfair to the Child?" *Hastings Center Report* 24 (November-December 1994): 363–369.

47. Lori Andrews et al., *Assessing Genetic Risks: Implications for Health and Social Policy* (Washington, DC: National Academy of Sciences, 1994).

48. See Holtzman and Marteau, "Will Genetics Revolutionize Medicine?" and Hubbard and Wald, *Exploding the Gene Myth*, 111 ff.

49. "Upside Risks: Social Consequences of Beneficial Biotechnology," in Carl Cranor, ed., *Are Genes Us? The Social Consequences of the New Genetics* (New Brunswick, NJ: Rutgers University Press, 1994): 155–179, at 157.

Chapter 2

1. John Martin Fischer and Mark Ravizza develop and defend this conception of responsibility in *Responsibility and Control: A Theory of Moral Responsibility* (New York: Cambridge University Press, 1998).

2. Ibid., chap. 1. The basic idea behind these two conditions is traceable to Aristotle, *Nicomachean Ethics*, Book III. See *The Complete Works of Aristotle*, Volumes I and II, Jonathan Barnes, trans. and ed. (Princeton, NJ: Princeton University Press, 1984).

3. Andrews et al., *Assessing Genetic Risks*, 4, and British Medical Association, *Human Genetics*, 34.

4. Clark, *The New Healers*, 54.

5. J. Mitchell et al., "What Young People Think and Do When the Option for Cystic Fibrosis Carrier Testing Is Available," *Journal of Medical Genetics* 30 (1993): 538–542.

6. See, for example, Juha Raikka, "Freedom and the Right (Not) to Know," *Bioethics* 12 (1998): 49–63.

7. Rosamond Rhodes defends this idea in "Genetic Links, Family Ties, and Social Bonds: Rights and Responsibility in the Face of Genetic Knowledge," *Journal of Medicine and Philosophy* 23 (February 1998): 10–30.

8. Lynn Hartman et al., "Efficacy of Bilateral Prophylactic Mastectomy in Women with a Family History of Breast Cancer," *New England Journal of Medicine* 340 (January 14, 1999): 77–84, and Eli Ginzberg, "Putting the Risk of Breast Cancer in Perspective," Ibid.: 141–146. In a more recent study, 143 women at increased risk of breast cancer were offered a prophylactic mastectomy, and 79 chose to have the surgery. At the beginning of the study, many of the women reported high levels of anxiety and other psychological problems. Following the surgery, however, these problems resolved. See M. B. Hatcher et al., "The Psychological Impact of Bilateral Prophylactic Mastectomy: Perspective Study Using Questionaires and Structural Interviews," *British Medical Journal* 322 (January 6, 2001): 76–79. Equally significant, Hanne Meijers-Heijboer et al. conducted a study showing that prophylactic bilateral total mastectomy substantially reduces the incidence of breast cancer among women with the

BRCA1 or BRCA2 mutation. "Breast Cancer after Prophylactic Bilateral Mastectomy in Women with BRCA1 or BRCA2 Mutations," *New England Journal of Medicine* 345 (July 19, 2001): 159–164.

9. Jonathan Berkowitz and Jack Snyder, "Racism and Sexism in Medically Assisted Conception," *Bioethics* 12 (1998): 25–44; and C. L. Ten, "The Use of Reproductive Technologies in Selecting the Sexual Orientation, Race, and the Sex of Children," Ibid.: 45–48.

10. See R. D. Terry et al., eds., *Alzheimer Disease* (Baltimore, MD: Lippincott, Williams and Wilkins, 1999); and A. E. Lang and A. M. Lozano, "Parkinson's Disease," *New England Journal of Medicine* 339 (October 8, 1998): 1044–1050; and Jeffrey Cummings, "Understanding Parkinson's Disease," Ibid.: 376–378.

11. Amos Shapira, "Wrongful Life Lawsuits for Faulty Genetic Counseling: Should the Impaired Newborn Be Entitled to Sue?" *Journal of Medical Ethics* 24 (1998): 369–375. Also, Brock, "The Non-Identity Problem and Genetic Harm: The Case of Wrongful Handicaps," *Bioethics* 9 (1995): 269–276; Buchanan et al., *From Chance to Choice*, chap. 6; Jeff McMahan, "Wrongful Life: Paradoxes in the Morality of Causing People to Exist," in Jules Coleman and Christopher Morris, eds., *Rational Commitment and Social Justice: Essays for Gregory Kavka* (Cambridge: Cambridge University Press, 1998): 208–247; Joel Feinberg, "Wrongful Life and the Counterfactual Element in Harming," in *Freedom and Fulfillment* (Princeton, NJ: Princeton University Press, 1992): 3–36; Heyd, *Genethics*, chap. 1; and Steinbock and McClamrock, "When Is Birth Unfair to the Child?"

12. In "Beware! Preimplantation Genetic Diagnosis May Solve Some Old Problems but It Also Raises New Ones," *Journal of Medical Ethics* 25 (1999): 114–120, bioethicists Heather Draper and Ruth Chadwick present a hypothetical case in which two deaf parents deliberately choose an embryo with congenital deafness for implantation. This would be difficult to justify, since the parents would be intentionally foreclosing opportunities for a future child.

13. This thesis derives from Jan Narveson's claim that we do not have a moral duty to make happy people, but only to make people happy. He argues that the benefit of an act is the good it brings to already existing people and does not include the good of people who come into existence as a result of the act in "Utilitarianism and New Generations," *Mind* 76 (1967): 62–72, and "Moral Problems of Population," *The Monist* 57 (1973): 62–86. John Broome and Adam Morton discuss different aspects of the moral asymmetry thesis in "The Value of a Person," *Proceedings of the Aristotelian Society*, Supplementary Volume 68 (1994): 167–198.

14. Joel Feinberg, in *Harm to Others*; Broome; Ibid.; Parfit, in *Reasons and Persons*; and Harris, in *Clones, Genes, and Immortality,* all define harm in comparative terms. That is, a person is harmed when she is made worse off than she would have been otherwise with respect to her interests. I avoid using the comparative sense of harm when considering whether being caused to exist with disabilities harms people, because these people would not exist without the disabilities they have, and a coherent comparison can be made only between two states of existence. The comparative sense of harm can be invoked only insofar as people exist and have interests. Otherwise, we should use an impersonal sense of harm, comparing two distinct potential lives of two distinct potential people.

15. Parfit defends this principle in *Reasons and Persons,* chapt. 18, and in "Comments," *Ethics* 96 (1986): 858 ff., as does Jonathan Glover, "Future People, Disability, and Screening," in P. Laslett and J. Fishkin, eds., *Justice Between Age Groups and Generations* (New Haven, CT: Yale University Press, 1992): 127–143.

16. This idea is discussed by Parfit in *Reasons and Persons* and by Harris in *Clones, Genes, and Immortality.*

17. *Reasons and Persons,* 488–489.

18. In "The Paradox of Future Individuals," *Philosophy & Public Affairs* 11 (1982): 93–112, Kavka defines a restricted life as "one that is significantly deficient in one or more of the major respects that generally make human lives valuable and worth living" (105). Yet Kavka further states that "restricted lives typically will be worth living, on the whole, for those who live them" (105). When I say that a life is not worth living for a person, I mean only a *severely* restricted life. McMahan offers insightful discussion of this and related issues in "Wrongful Life: Paradoxes in the Morality of Causing People to Exist," and in "Cognitive Disability, Misfortune, and Justice," *Philosophy & Public Affairs* 25 (1996): 3–35.

19. *Reasons and Persons,* chap. 16, and "Comments," 854–862. Others who address this problem include McMahan, "Cognitive Disability" and "Wrongful Life"; Kavka, "The Paradox of Future Individuals"; James Woodward, "The Non-Identity Problem," *Ethics* 96 (1986): 804–831; Matthew Hanser, "Harming Future People," *Philosophy & Public Affairs* 19 (1990): 47–70; Heyd, *Genethics,* chaps. 4 and 6; Brock, "The Non-Identity Problem and Genetic Harm"; and Buchanan et al., *From Chance to Choice,* chap. 6. Robert Adams introduced the Non-Identity Problem in "Existence, Self-Interest, and the Problem of Evil," *Nous* 13 (1979): 65–76.

20. "Future People, Disability, and Screening," 141.

21. Ibid., 142.

22. Here I assume that the same number of people will exist in the different outcomes. This avoids complications involving different numbers of people and having to determine which group is better or worse off than others. See Parfit's discussion of "Same People Choices," "Same Number Choices," and "Different Number Choices" in *Reasons and Persons*, 356 ff, and Heyd, *Genethics*.

23. "Wrongful Life and the Counterfactual Element in Harming."

24. *Genethics*, chap. 2.

25. Helga Kuhse and Peter Singer claim that "we can, of course, damage the embryo in such a way as to cause harm to the sentient being it will become, if it lives; but if it never becomes a sentient being, the embryo has not been harmed." From "Individuals, Humans, and Persons: The Issue of Moral Status," in *Embryo Experimentation*, 82. Furthermore, Harris says that "harm done at the pre-person (embryo) stage will be harm done to the actual person *she* becomes. It is a form of delayed-action wrongdoing." *Clones, Genes, and Immortality*, 153. As I have been arguing, persons can be harmed by what we do or fail to do to embryos even if embryos do not strictly speaking *become* persons.

26. Kitcher provides a helpful overview and discussion of the genetic causes of these and other diseases in *The Lives to Come*.

27. See Daniels, "The Genome Project, Individual Differences, and Just Health Care"; Buchanan, "Equal Opportunity and Genetic Intervention," *Social Philosophy & Policy* 12 (1995): 105–135; and "Choosing Who Will Be Disabled: Genetic Intervention and the Morality of Inclusion," Ibid., 13 (1996): 18–46; and Buchanan et al., *From Chance to Choice*, chap. 3.

28. Silver discusses different aspects of embryo selection in *Remaking Eden*, 267 ff.

29. R.G. Edwards and J. Purdy, *Human Conception in Vitro* (London: Academic Press, 1981), 373. See also Bonnie Steinbock, *Life Before Birth* (New York: Oxford University Press, 1992), and Mary Ann Warren, *Moral Status* (Oxford: Clarendon Press, 1998).

30. Kitcher offers a similar model to measure quality of life in *The Lives to Come*, as does Brock, "Quality of Life Measures in Health Care and Medical Ethics," in *Life and Death: Philosophical Essays in Biomedical Ethics* (New York: Cambridge University Press, 1993): 268–324.

31. See Nagel, "Death," in *Mortal Questions*, 1–10, and Parfit, *Reasons and Persons*, 165–186.

32. Michael Stocker formulates the issue in this way in "Parfit and the Time of Value," in J. Dancy, ed., *Reading Parfit* (Oxford: Blackwell, 1997): 54–70, at

65. Thomas Hurka presents a model that measures the quality of a person's life as a whole in terms of averaging achievements in earlier and later stages of that life in *Perfectionism* (Oxford: Oxford University Press, 1993), 70 ff.

33. See Terry et al., *Alzheimer Disease*.

34. Consider G.H. Hardy's account of the Indian mathematician Ramanujan, who died at an early age: "The real tragedy about Ramanujan was not his early death. It is of course a disaster that any great man should die young, but a mathematician is often comparatively old at thirty, and his death may be less of a catastrophe than it seems." *Ramanujan: Twelve Letters on Subjects Suggested by His Life and Work* (Cambridge: Cambridge University Press, 1940), 6. Commenting on this point, Hurka writes that "if a life, for example, has already had its best moments, nothing now or in the future can alter its aggregate worth, and it does not matter whether it is added to or not," *Perfectionism*, 76. It is instructive to consider A. E. Housman's poem "To An Athlete Dying Young" in this regard as well.

35. See Lewis Rowland and Neil Shneider, "Amyotrophic Lateral Sclerosis," *New England Journal of Medicine* 344 (May 31, 2001): 1688–1700.

36. This definition of moral requirements and moral permissions closely follows that of Shelly Kagan in *The Limits of Morality* (Oxford: Clarendon Press, 1989): 65–66. See also Samuel Scheffler, *Human Morality* (New York: Oxford University Press, 1992): 17–22.

37. Kuhse offers a sustained critical analysis of this principle in *The Sanctity-of-Life Doctrine in Medicine: A Critique* (Oxford: Clarendon Press, 1987).

38 John Rawls, *A Theory of Justice* (Cambridge, MA: Harvard Belknap Press, 1971): 1–27, 78–83, 152–153, and "Social Unity and Primary Goods," in A. Sen and B. Williams, eds., *Utilitarianism and Beyond* (Cambridge: Cambridge University Press, 1982): 159–185; Nagel, *Equality and Partiality* (Oxford: Oxford University Press, 1991); Larry Temkin, *Inequality* (Oxford: Oxford University Press, 1993); Parfit, "Equality or Priority?"; Lindley Lecture (Lawrence, KS: University of Kansas Press, 1995); T.M. Scanlon, "Contractualism and Utilitarianism," in Sen and Williams, 103–128; and *What We Owe to Each Other* (Cambridge, MA: Harvard University Press, 1998); Dworkin, *Sovereign Virtue*, chap. 1, and Buchanan et al., *From Chance to Choice*, chap. 3.

39. *A Theory of Justice*, 83, 152–153.

40. Parfit, *Reasons and Persons*, 428–436.

41. Parfit, Ibid., Nagel, *Equality and Partiality*, 73; Scanlon, "Contractualism and Utilitarianism," and *What We Owe to Each Other*; and Kamm, *Morality, Mortality*, Volume 1. Kamm says that "we could minimally change a strict maximin theory in the following way: only when what would be gained by the worst

off is a gain that makes a real value significant difference to him should we forego giving the better-off person a much greater gain" (270). She calls this the "modified maximin rule." We can interpret "value significant difference" in terms of changes in people's health status measured against an absolute baseline of a decent minimum of health.

42. These figures, as well as those for the incidence of other genetic disorders, are given by Mehlman and Botkin in *Access to the Genome*, 41.

43. "Choosing Who Will Be Disabled," 32. Emphases added. Also, Buchanan et al., *From Chance to Choice*, chap. 7.

44. Prenatal Diagnosis and Discrimination Against the Disabled." *Journal of Medical Ethics* 25 (1999): 163–171, at 87. What Gillam says about fetuses applies to embryos as well.

45. "Choosing Who Will Be Disabled," 33.

46. See Gillam's discussion of support for people with thalassemia in Greece in "Prenatal Diagnosis," 84–85.

47. "Equality," in *Mortal Questions*, 106–127.

48. *Remaking Eden*, 9 ff.

Chapter 3

1. See Daniels, *Just Health Care*, Daniels and Sabin, "Determining 'Medical Necessity' in Mental Health Practice," and Buchanan et al., *From Chance to Choice*, chap. 1 and 7.

2. Leroy Walters and Julie Gage Palmer make this mistake in *The Ethics of Human Gene Therapy* (New York: Oxford University Press, 1997), chap. 3.

3. In addition to the articles on this issue which I mentioned in chapter 1, see Ingmar Persson, "Genetic Therapy, Identity, and the Person-Regarding Reasons," *Bioethics* 9 (1995): 16–31, and Robert Elliot, "Gene Therapy, Person-Regarding Reasons, and the Determination of Identity," Ibid., 11 (1997): 151–160.

4. Patricia Baird provides an excellent overview of these issues in "Altering Human Genes: Social, Ethical, and Legal Implications," *Perspectives in Biology and Medicine* 37 (1994): 566–575. Also, Walters and Palmer, *The Ethics of Human Gene Therapy*.

5. Walters and Palmer, Ibid., 73.

6. Hubbard and Wald discuss this problem in *Exploding the Gene Myth*, 108–116, as does Clark in *The New Healers*, chap. 8. More positive views of gene therapy are as follows: R. M. Blaese et al., "Treatment of Severe Combined Immunodeficiency (SCID) Due to Adenosine Deaminase Deficiency

with CD34+ Selected Autologous Blood Cells Transduced with a Human ADA Gene," *Human Gene Therapy* 4 (1993): 521–527; M. Grossman et al., "Successful Ex Vivo Gene Therapy Directed to Liver in a Patient with Familial Hypercholesterolemia," *Nature Genetics* 6 (1994): 335–341; R. C. Boucher et al., "Gene Therapy for Cystic Fibrosis Using E4 Deleted Adenovirus: A Trial in the Nasal Cavity," *Human Gene Therapy* 5 (1994): 615–639; Melissa Rosenfeld, "Human Artificial Chromosomes Get Real," *Nature Genetics* 15 (1997): 333–335; and Eugene Kaji and Jeffrey Leiden, "Gene and Stem Cell Therapies," *Journal of the American Medical Association* 285 (February 7, 2001): 545–550.

7. Sheryl Stolberg, "Youth's Death Shakes New Field of Gene Experiments on Humans," *New York Times*, January 27, 2000, D1–D2.

8. Natalie Southworth, "Toronto Death Raises Questions About the Risks of Gene Therapy," *Globe and Mail*, March 6, 2000, A3.

9. Sheryl Stolberg, "A Second Death Linked to Gene Therapy," *New York Times*, May 4, 2000, D1. Eliot Marshall points out that the lack of an ideal vector for delivering genes into targeted cells remains the central problem in "Gene Therapy on Trial," *Science* 288 (May 12, 2000): 951–957.

10. Clark proposes liposomes as alternative delivery vectors for genes in *The New Healers*, 120.

11. See Thomas Bumol and August Watanabe, "Genetic Information, Genomic Technologies, and the Future of Drug Discovery," *Journal of the American Medical Association* 285 (February 7, 2001): 551–555.

12. D. A. Roth et al., "Nonviral Transfer of the Gene Encoding Coagulation Factor VIII in Patients with Severe Hemophilia A," *New England Journal of Medicine* 344 (June 7, 2001): 1735–1742.

13. Meredith Wadman, "Gene Therapy Pushes On, Despite Doubts," *Nature* 397 (January 14, 1999): 94.

14. W. French Anderson gives a more favorable assessment of this issue in "Human Gene Therapy: Why Draw a Line?" *Journal of Medicine and Philosophy* 14 (December 1989): 681–693.

15. "Risks Inherent in Fetal Gene Therapy," *Nature* 397 (February 4, 1999), 383.

16. Ya-Ping Tang et al., "Genetic Enhancement of Learning and Memory in Mice," *Nature* 401 (September 2, 1999): 63–69. Depending on the level of one's cognitive functioning, this type of genetic intervention could be labeled as either "therapy" or "enhancement."

17. See McMahan, "Cognitive Disability, Misfortune, and Justice"; Buchanan, "Choosing Who Will Be Disabled: Genetic Intervention and the Morality of Inclusion"; and Buchanan, et al., *From Chance to Choice*, chap. 7.

18. Rawls, for one, criticizes utilitarianism for this reason in *A Theory of Justice*, 27 ff.

19. See, for example, G. Q. Maguire and Ellen McGee, "Implantable Brain Chips? Time for Debate," *Hastings Center Report* 29 (January-February 1999): 7–13.

20. *Access to the Genome*, 124–128.

21. Ibid., 125.

22. These rates of incidence (though not the cost of treatment) are given by Mehlman and Botkin, Ibid., 41.

23. See Clark, *The New Healers*, chap. 9, "Gene Therapy for Cancer," 134–160.

24. Jon Gordon, "Genetic Enhancement in Humans," *Science* 283 (March 26, 1999): 2023–2024.

25. Eric Juengst, "Can Enhancement Be Distinguished from Prevention in Genetic Medicine?" *Journal of Medicine and Philosophy* 22 (1997): 125–142, and "What Does Enhancement Mean?" in Erik Parens, ed., *Enhancing Human Traits: Ethical and Social Implications* (Washington, DC: Georgetown University Press, 1998): 27–47, at 27. Also, Dan Brock, "Enhancements of Human Function: Some Distinctions for Policymakers," Ibid., 48–69.

26. *The Ethics of Human Gene Therapy*, 110. Instead of distinguishing between treatments and enhancements, Walters and Palmer distinguish between health-related and non–health-related enhancements. But I do not find this distinction to be very helpful.

27. Brock points this out in "Enhancements of Human Function," 59. Marc Lappe makes a more compelling case for the same point in *The Tao of Immunology*.

28. Kavka develops and defends the idea that competitive goods are continuous in "Upside Risks," 164–165.

29. Kavka, "Upside Risks," 167. Also, Brock, "Enhancements of Human Function," 60; and Buchanan et al., *From Chance to Choice*, chap. 8.

30. Rawls makes this point in *A Theory of Justice*, 7–11, and in "Social Unity and Primary Goods," 162. See also Daniels, *Just Health Care*.

31. Walters and Palmer present this thought-experiment in *The Ethics of Human Gene Therapy*, 123–128. As they note, Jonathan Glover introduced this idea in *What Sort of People Should There Be?* (Harmondsworth: Penguin, 1984).

32. Drawing on the work of Lionel trilling and Charles Taylor, Carl Elliott discusses cognitive and affective enhancements that undermine what he calls the "ethics of authenticity" in "The Tyranny of Happiness: Ethics and Cosmetic Psychopharmacology," in Parens, *Enhancing Human Traits*, 177–188. Also relevant to this issue is Harry Frankfurt, "Identification and Externality," in Frankfurt, *The Importance of What We Care About* (New York: Cambridge University Press, 1989): 58–68.

33. David and Clare Galton, "Francis Galton: And Eugenics Today," *Journal of Medical Ethics* 24 (1998): 99–105. Also, Buchanan et al., *From Chance to Choice*, chap. 2.

34. See, for example, Daniel Kevles, *In the Name of Eugenics: Genetics and the Uses of Human Heredity* (New York: Alfred A. Knopf, 1985), and Troy Duster, *Backdoor to Eugenics* (New York: Routledge, 1990). Cf. John Wagner, "Gene Therapy Is Not Eugenics," *Nature Genetics* 15 (1997), 234.

35. This term originally was coined by Joshua Lederberg in "Molecular Biology, Eugenics, and Euphenics," *Nature* 198 (1963): 428–429. James Neel elaborates further on this idea in "Looking Ahead: Some Genetic Issues of the Future," *Perspectives in Biology and Medicine* 40 (1997): 328–347.

36. "Which Slopes Are Slippery?" in Williams, *Making Sense of Humanity* (Cambridge: Cambridge University Press, 1995): 213–223, at 213.

37. Ibid., 213.

38. See Trudy Govier, "What's Wrong with Slippery Slope Arguments?" *Canadian Journal of Philosophy* 12 (June 1982): 303–316. Also, Wibren van den Berg, "The Slippery Slope Argument," *Ethics* 102 (October 1991): 42–65.

39. Here I follow Govier's formulation, 311.

40. See E. Berger and B. Gert, "Genetic Disorders and the Ethical Status of Germ-Line Gene Therapy," *Journal of Medicine and Philosophy* 16 (1991): 667–683. Also David Resnick, "Debunking the Slippery Slope Argument About Human Germ-Line Gene Therapy," Ibid., 19 (1994): 23–40.

41. *The Lives to Come*, 202. See also Glover, *What Sort of People Should There Be?*

Chapter 4

1. Ian Wilmut et al., "Viable Offspring Derived from Fetal and Adult Mammalian Cells," *Nature* 385 (February 27, 1997): 310–313.

2. *Cloning Human Beings: Report and Recommendations of the National Bioethics Advisory Commission* (Rockville, MD, 1997). James Childress, Susan Wolf, Courtney Campbell, Daniel Callahan, and Erick Parens discuss different moral

and legal aspects of the Report in "Cloning Human Beings: Responding to the National Bioethics Advisory Commission's Report," *Hastings Center Report* 27 (September–October 1997): 9–22.

3. Noteworthy critiques of cloning on these grounds are offered by Leon Kass, "The Wisdom of Repugnance: Why We Should Ban the Cloning of Humans," *The New Republic*, June 2, 1997: 17–26; George Annas, "Why We Should Ban Human Cloning," *New England Journal of Medicine* 339 (July 9, 1998): 122–125; and Soren Holm, "A Life in the Shadow: One Reason Why We Should Not Clone Humans," *Cambridge Quarterly of Healthcare Ethics* 7 (1998): 160–162. I address specific claims by Annas and Kass later in this chapter.

4. John Harris argues along these lines in "Cloning and Human Dignity," *Cambridge Quarterly of Healthcare Ethics* 7 (1998): 163–167, and *Clones, Genes, and Immortality*. See also Gregory Pence, *Who's Afraid of Human Cloning* (Lanham, MD: Rowman & Littlefield, 1998), and Martha Nussbaum and Cass Sunstein, eds., *Clones and Clones: Facts and Fantasies About Human Cloning* (New York: W. W. Norton, 1999).

5. "Ethical Aspects of Genetic Controls," *New England Journal of Medicine* 285 (1971): 776–783, and *The Ethics of Genetic Control: Ending Reproductive Roulette* (Garden City, NY: Anchor Press, 1974).

6. This is the second formulation of the categorical imperative. The first formulation says: "I ought never to act in such a way that I could not also will that my maxim should be a universal law." From the *Foundations of the Metaphysics of Morals* (1785), trans. L. W. Beck, second edition (New York: Macmillan, 1990), 429 for the second formulation of the categorical imperative, 402 for the first. Page references are to the Royal Prussian Academy edition. F. M. Kamm offers a contemporary defense of the Kantian notion of a person as an inviolable end in itself in "Nonconsequentialism, the Person as an End-in-Itself, and the Significance of Status," and in *Morality, Mortality, Volume II: Rights, Duties, and Status* (Oxford: Oxford University Press, 1996).

7. Patricia Baird discusses the two types of cloning in "Cloning of Animals and Humans: What Should the Policy Response Be?" *Perspectives in Biology and Medicine* 42 (Winter 1999): 179–194. In her conclusion, Baird writes, "to use a somatic-cell cloning technique to allow an infertile couple to have a child doesn't necessarily offend the Kantian principle, but it does breach a natural barrier, which once passed, leaves us with no clear place to stop" (192). This suggests a slippery slope from using cloning as a form of assisted reproduction to more morally questionable uses. But, as I pointed out in chapter 3 and will reiterate later in this chapter, slippery-slope arguments are too weak to sup-

port any claims about one thing leading to or causing another. This includes claims about cloning. If properly formulated and implemented, public policy and legislation should be able to clearly demarcate using cloning for disease prevention or treatment from using it for physical or mental enhancement. There need not be a slippery slope from one use to the other.

8. "Full-Term Development of Mice from Enucleated Oocytes Injected with Cumulus Cell Nuclei," *Nature* 394 (July 23, 1998): 369–374. Davor Solter argues that the result from the work of Wakayama et al. shows that doubts about cloning in the wake of Wilmut's result can be set aside in his accompanying article, "Dolly *Is* a Clone—and No Longer Alone," Ibid., 315–316.

9. Michael Schuman, "Korean Experiment Fuels Cloning Debate: More Work Is Needed to Prove a Live Birth Is Possible," *Wall Street Journal*, December 21, 1998.

10. This is known as the "Hayflick Limit," named after Leonard Hayflick, who proposed and defended the theory of programmed cell death in "The Cellular Basis for Biological Aging," in Hayflick and C. E. Finch, eds., *Handbook of the Biology of Aging* (New York: Van Nostrand, 1977): 159–186. I discuss the biological mechanisms of aging in more detail in chapter 5.

11. See, for example, D. Broccoli and H. Cooke, "Aging, Healing, and the Metabolism of Telomeres," *American Journal of Human Genetics* 52 (1993): 657–660, and B. W. Stewart, "Mechanisms of Apoptosis: Integration of Genetic, Biochemical, and Cellular Indicators," *Journal of the National Cancer Institute* 86 (1994): 1286–1296.

12. Paul Shiels et al., "Analysis of Telomere Length in Cloned Sheep," *Nature* 399 (May 27, 1999): 316–317.

13. See Richard Michod, "What Good Is Sex?" in *The Sciences*, special issue: "The Promise and Peril of Cloning," 37 (September-October 1997): 42–46.

14. J. Arjan et al. point this out in "Diminishing Returns from Mutation Supply Rate in Asexual Populations," *Science* 283 (January 15, 1999): 404–406, as do Adam Eyre-Walker and Peter Keightley, in "High Genomic Deleterious Mutation Rates in Hominids," *Nature* 397 (January 28, 1999): 344–347.

15. Silver, *Remaking Eden*, 121.

16. Ibid., 121.

17. "Would Cloned Humans Really Be Like Sheep?" *New England Journal of Medicine* 340 (February 11, 1999): 471–475, at 473.

18. This idea derives from Derek Parfit's discussion of the "Social Discount Rate" in *Reasons and Persons*, 480–486. I examine this idea in more detail in chapter 5.

19. Michod, "What Good Is Sex?" 43. Also, Eyre-Walker and Keightley.

20. "Cloning Human Beings: An Assessment of the Ethical Issues Pro and Con," in John Arras and Bonnie Steinbock, eds., *Ethical Issues in Modern Medicine*, Fifth Edition (Mountain View, CA: Mayfield, 1999): 484–496, at 495.

21. Kant explains the intrinsic worth of persons in these terms in the *Foundations*, 394 ff. In "Goodbye Dolly?: The Ethics of Human Cloning," *Journal of Medical Ethics* 23 (1997): 323–329, Harris questions how helpful the Kantian distinction between means and ends can be in debating the moral issues regarding cloning. But my discussion and the examples I use show that it is the most helpful method of examining these issues.

22. Kitcher discuses this case in the context of cloning in "Whose Self Is It, Anyway?" *The Sciences*, "Promise and Peril," 37 (1997): 58–62.

23. Denise Grady, "Baby Conceived to Provide Cell Transplant for His Dying Sister," *New York Times*, October 4, 2000.

24. "Why We Should Ban Human Cloning," 123.

25. Silver, *Remaking Eden*, 195.

26. *Eisenstadt v. Baird*, 405 U.S. 438, 453 (1972). This case is cited and analyzed by Lori Andrews in "Mom, Dad, Clone: Implications for Reproductive Privacy," *Cambridge Quarterly of Healthcare Ethics* 7 (Spring 1998): 176–186. It should be noted that Andrews argues, on grounds different from the legal one at issue here, that cloning should not be permitted.

27. For discussion of the distinction between negative and positive rights, see, for example, Kamm, *Morality, Mortality, Volume II: Rights, Duties, and Status.* Judith Jarvis Thomson, *The Realm of Rights* (Cambridge, MA: Harvard University Press, 1990).

28. "Why We Should Ban Human Cloning," 123.

29. "Genes, Environment, and Personality," *Science* 246 (June 17, 1994): 1700–1701.

30. "Whenever the Twain Shall Meet," in *The Sciences* 37 (; "Promise and Peril," 52–54, at 54.

31. "The Child's Right to an Open Future," in W. Aiken and H. LaFollette, eds., *Whose Child?: Children's Rights, Parental Authority, and State Power* (Totowa, NJ: Rowman & Littlefield, 1980): 124–153.

32. See, for example, the series of articles in "The Promise of Tissue Engineering," *Scientific American*, April 1999, especially David Mooney and Antonios Mikos, "Growing New Organs," 60–65. There is a dark side to this technology as well. Andrews and Dorothy Nelkin explore how tissues, cells, and genes can be harvested and turned into commodities on the open market in

Body Bazaar: The Market for Human Tissue in the Biotechnology Age (New York: Crown Publishers, 2001).

33. Ibid., 63.

34. See Ronald Hart et al., "Born Again?" *The Sciences* 37 (1997), "Promise and Peril," 47–51.

35. Reported in the London *Sunday Times*, October 19, 1997.

36. See Steve Mirsky and John Rennie, "What Cloning Means for Gene Therapy," *Scientific American*, June 1997: 122–123.

37. "The Wisdom of Repugnance," 24.

Chapter 5

1. See, for example, Gregory Armstrong et al., "Trends in Infectious Diseases Mortality in the United States during the 20th Century," *Journal of the American Medical Association* 281 (January 6, 1999): 61–66, and John Cairns, *Matters of Life and Death: Perspectives on Public Health, Molecular Biology, Cancer and the Prospects for the Human Race* (Princeton: Princeton University Press, 1997).

2. P. B. Medawar and J. S. Medawar, *The Life Science: Current Ideas of Biology* (New York: Harper & Row, 1977), especially chapter 20, "Senescence"; Caleb Finch, *Longevity, Senescence and the Genome* (Chicago: University of Chicago Press, 1990); Michael Rose, *Evolutionary Biology of Aging* (Oxford: Oxford University Press, 1991); and S. Jay Olshansky and Bruce Carnes, *The Quest for Immortality: Science at the Frontiers of Aging* (New York: Norton, 2001).

3. Andrea Bodner et al., "Extension of Life-Span by Introduction of Telomerase into Normal Human Cells," *Science* 279 (January 16, 1998): 349–352.

4. At the Geron Corporation in Menlo Park, CA, as reported by Nicholas Wade, "Immortality, of a Sort, Beckons Biologists," *New York Times*, November 17, 1998. Gordon Keller and Ralph Snodgrass explore the biological implications of this in "Human Embryonic Stem Cells: The Future Is Now," *Nature Medicine* 5 (February 1999): 151–152. Some of the ethical and policy issues surrounding ES cell research are discussed by Karen Lebacqz and the Geron Ethics Advisory Board, "Research with Human Embryonic Stem Cells: Ethical Considerations," *Hastings Center Report* 29 (1999): 31–36.

5. See the series of articles in the special section on apoptosis in *Science* 281 (August 28, 1998): 1301–1325, especially Gerard Evan and Trevor Littlewood, "A Matter of Life and Cell Death," 1317–1325. See also William Clark, *At War Within: The Double-Edged Sword of the Immune System* (Oxford: Oxford University Press, 1995).

6. Marcia Baringa, "Is Apoptosis Key in Alzheimer's Disease?" *Science* 281, 1303–1304.

7. Xu-Rong Jiang et al., "Telomerase Expression in Human Somatic Cells Does Not Induce Changes Associated with a Transformed Phenotype," *Nature Genetics* 21 (January 1999): 111–114.

8. D. Promislow, "Longevity and the Barren Aristocrat," *Nature* 396 (December 24–31, 1998): 719–720. Also, R. Westendorp and T. Kirkwood, "Human Longevity at the Cost of Reproductive Success," Ibid., 743–744.

9. George Williams, "Pleiotropy, Natural Selection, and the Evolution of Senescence," *Evolution* 11(1957): 398–411. This hypothesis is elaborated further by Nesse and Williams in *Why We Get Sick: The New Science of Darwinian Medicine*, and "Evolution and the Origins of Disease." See also Ewald, *Evolution of Infectious Disease*.

10. "Evolution and the Origins of Disease," 86. As mentioned in chapter 1, this idea derives from Theodosius Dobzhansky.

11. Gerald Pier et al., "Salmonella Typhi Uses CFTR to Enter Intestinal Epithelial Cells," *Nature* 393 (May 20, 1998): 79–82.

12. *Why We Get Sick*, chap. 8, and "Evolution and the Origins of Disease," 91–93.

13. *Reasons and Persons*, 486 ff.

14. Ibid., 486.

15. "Evolution and the Origins of Disease," 92.

16. Russell Ross, "Atherosclerosis—An Inflammatory Disease," *New England Journal of Medicine* 340 (January 14, 1999): 115–123.

17. "Evolution and the Origins of Disease," 92.

18. See *Reasons and Persons*, Appendix F.

19. *Genethics: Moral Issues in the Creation of People*, 94 ff.

20. Kamm employs this principle in her analysis of allocating scarce medical resources in *Morality, Mortality: Volume 1: Death and Whom to Save From It* (Oxford: Oxford University Press, 1993), part III.

21. See Thomas Hurka, "Value and Population Size," *Ethics* 93 (April 1983): 134–147, and Parfit, *Reasons and Persons*, chap. 17. For discussions of the issue of fairness as it pertains to the young supporting the old, see the essays in Peter Laslett and James Fishkin, eds., *Justice Between Age Groups and Generations* (New Haven: Yale University Press, 1992).

22. See Cairns, *Matters of Life and Death*, chap. 6, and Garrett Hardin, *Living Within Limits* (New York: Oxford University Press, 1995).

23. "Upside Risks," 162.

24. This idea is defended by Daniels in his prudential life span account to determine a just distribution of health care, in *Just Health Care*, and *Am I My Parents' Keeper?* (New York: Oxford University Press, 1988).

25. See, for example, Dennis McKerlie, "Equality Between Age Groups," *Philosophy & Public Affairs* 22 (Summer 1992): 178–198.

26. "Aging and the Allocation of Resources," *American Journal of Ethics & Medicine*, 1998: 4–9. Callahan offers a more sustained discussion of this problem in *Setting Limits* (New York: Simon & Schuster, 1987), and *What Kind of Life?* (New York: Simon & Schuster, 1990).

27. Including the interests of our descendants within our prudential interests would generate moral reasons for not manipulating the genetic mechanisms of aging. Since the manipulation would be done at the germ line, any adverse effects on other genes regulating cell growth and repair would be passed on to these descendants.

28. This idea derives from James Fries's important papers, "Aging, Natural Death, and the Compression of Morbidity," *New England Journal of Medicine* 322 (July 17, 1980): 130–135, and "The Compression of Morbidity," *Milbank Memorial Fund Quarterly* 61 (1983): 347–419.

29. *Reasons and Persons*, 245 ff. Also, Parfit, "Lewis, Perry, and What Matters," in Rorty, 91–107.

30. Jonathan Swift, *Gulliver's Travels* (1726) (London: Penguin, 1941).

31. See Bernard Williams's discussion of the distinction between categorical and conditional desires to go on living in "The Makropulos Case: Reflections on the Tedium of Immortality," in *Problems of the Self* (Cambridge: Cambridge University Press, 1973): 82–100.

32. *Perfectionism*, 72.

33. I am grateful to Jeff McMahan for raising this issue, and for many other helpful comments on this section of this chapter.

34. *Reasons and Persons*, 206.

35. "Survival and Identity," 30.

36. *Reasons and Persons*, 313.

37. Ibid., 337.

38. See Ted Abel et al., "Memory Suppressor Genes: Inhibitory Constraints on the Storage of Long-Term Memory," *Science* 279 (January 16, 1998): 338–341. Also, John Lisman and Justin Fallon, "What Maintains Memories?" *Science* 283 (January 15, 1999): 339–340.

39. In Jorge Luis Borges, *Ficciones*, Anthony Kerrigan, trans. and ed. (New York: Grove Press, 1962): 107–115.

40. This sentiment is expressed also by Hans Jonas, "The Burden and Blessing of Mortality," *Hastings Center Report* 22 (January-February 1992): 34–40, and by Leon Kass, "Mortality and Morality: The Virtues of Finitude," in *Toward a More Natural Science: Biology and Human Affairs*, 299–317. Also instructive in this regard is a passage from John Hick: "It is the very finitude of our earthly life, its haunting brevity, that gives it shape and value by making time precious and choice urgent. If we had before us an endless temporal vista, devoid of the pressure of an approaching end, our life would lose its present character as offering a continuum of choices, small and large, through which we participate in our own gradual creating. There is thus much to be said for the view that the formation of persons through their own freedom requires the boundaries of birth and death." From "A Possible Conception of Life After Death," in *Disputed Questions* (New Haven: Yale University Press, 1993): 183–196, at 189. This is not to deny the possibility of life after death as such, only to question the idea of an indefinite continuation of psychological life as we know it.

BIBLIOGRAPHY

Abel, Ted, et al. 1998. "Memory Suppressor Genes: Inhibitory Constraints on the Storage of Long-Term Memory." *Science* 279 (January 16): 338–341.

Adams, Robert. 1979. "Existence, Self-Interest, and the Problem of Evil." *Nous* 13: 65–76.

Anderson, W. French. 1989. "Human Gene Therapy: Why Draw a Line?" *Journal of Medicine and Philosophy* 14 (December): 681–693.

———. 1995. "Gene Therapy." *Scientific American*, September: 124–128.

———. 1999. "Risks Inherent in Fetal Gene Therapy." *Nature* 397 (February 4): 383.

———. 2000. "The Best of Times, the Worst of Times." *Science* 288 (April 28): 627–629.

Andreasen, Nancy. 1999. "Understanding the Causes of Schizophrenia." *New England Journal of Medicine* 340 (February 25): 645–647.

Andrews, Lori, et al. 1994. *Assessing Genetic Risks: Implications for Health and Social Policy* Washington, DC: National Academy of Sciences.

Andrews, Lori. 1998. "Mom, Dad, Clone: Implications for Reproductive Privacy." *Cambridge Quarterly of Healthcare Ethics* 7 (Spring): 176–186.

Andrews, Lori, and Nelkin, Dorothy. 2001. *Body Bazaar: The Market for Human Tissue in the Biotechnology Age.* New York: Crown Publishers.

Annas, George. 1998. "Why We Should Ban Human Cloning." *New England Journal of Medicine* 339 (July 9): 122–125.

Aristotle. 1984. *Nicomachean Ethics.* In *The Complete Works of Aristotle,* Volume II, Jonathan Barnes trans. and ed. Princeton, NJ: Princeton University Press.

Arjan, J., et al. 1999. "Diminishing Returns from Mutation Supply Rate in Asexual Populations." *Science* 283 (January 15): 404–406.

Armstrong, D. M. 1969. *A Materialist Theory of Mind.* Cambridge: Cambridge University Press.

Armstrong, Gregory, et al. 1999. "Trends in Infectious Disease Mortality in the United States During the 20th Century." *Journal of the American Medical Association* 281 (January 6): 61–66.

Austin, C. R. 1989. *Human Embryos: The Debate on Assisted Reproduction.* Oxford: Oxford University Press.

Baird, Patricia. 1994. "Altering Human Genes: Social, Ethical, and Legal Implications." *Perspectives in Biology and Medicine* 37 (Fall): 566–575.

———. 1999. "Cloning of Animals and Humans: What Should the Policy Response Be?" *Perspectives in Biology and Medicine* 42 (Winter): 179–194.

Baker, Lynn Rudder. 2000. *Persons and Bodies: A Constitution View.* New York: Cambridge University Press.

Barinaga, Marcia. 1998. "Is Apoptosis Key in Alzheimer's Disease?" *Science* 281: 1303–1304.

Berger, E., and Gert, B. 1991. "Genetic Disorders and the Ethical Status of Germ-Line Gene Therapy." *Journal of Medicine and Philosophy* 16 (December): 667–683.

Berkowtiz, Jonathan, and Snyder, Jack. 1998. "Racism and Sexism in Medically Assisted Conception." *Bioethics* 12 (January): 25–44.

Blaese, R. M., et al. 1993. "Treatment of Severe Combined Immunodeficiency (SCID) Due to Deaminase Deficiency with CD34+ Selected Autologous Blood Cells Transduced with a Human ADA Gene." *Human Gene Therapy* 4: 521–527.

———. 1995. "T Lymphocyte-Directed Gene Therapy for ADA-SCID: Initial Trial Results After Four Years." *Science* 270: 475.

Bodner, Andrea, et al. 1998. "Extension of Life-Span by Introduction of Telomerase into Normal Human Cells." *Science* 279 (January 16): 349–352.

Boorse, Christopher. 1977. "Health as a Theoretical Concept." *Philosophy of Science* 44: 542–571.

———. 1981. "On the Distinction Between Disease and Illness," in M. Cohen et al., eds., *Medicine and Moral Philosophy.* Princeton, NJ: Princeton University Press: 3–22.

Borges, Jorge Luis. 1962. *Ficciones,* Anthony Kerrigan, trans. and ed. New York: Grove Press.

Bouchard, Thomas. 1994. "Genes, Environment, and Personality." *Science* 247 (June 17): 1700–1701.

———. 1997. "Whenever the Twain Shall Meet." In *The Sciences* 37, Special Issue: "The Promise and Peril of Cloning": 52–54.

Boucher, R. C., et al. 1994. "Gene Therapy for Cystic Fibrosis Using E4 Deleted Adenovirus: A Trial in the Nasal Cavity." *Human Gene Therapy* 5: 615–639.

British Medical Association. 1998. *Human Genetics: Choice and Responsibility.* Oxford: Oxford University Press.

Broccoli, D., and Cooke, H. 1993. "Aging, Healing, and the Metabolism of Telomeres." *American Journal of Human Genetics* 52: 657–660.

Brock, Dan. 1993. "Quality of Life Measures in Health Care and Medical Ethics," in Brock, *Life and Death: Philosophical Essays in Biomedical Ethics*. New York: Cambridge University Press: 268–324.

———. 1995. "The Non-Identity Problem and Genetic Harm: The Case of Wrongful Handicaps." *Bioethics* 9 (July): 269–276.

———. 1998. "Entitlements of Human Function: Some Distinctions fo Policymakers," in Erik Parens, ed., *Enhancing Human Traits: Ethical and Social Implications*. Washington, DC: Georgetown University Press: 27–47.

———. 1999. "Cloning Human Beings: An Assessment of the Ethical Issues Pro and Con," in John Arras and Bonnie Steinbock, eds., *Ethical Issues in Modern Medicine*. Fifth Edition. Mountain View, CA: 484–496.

Broome, John. 1991. *Weighing Goods*. Oxford: Blackwell.

Broome, John, and Morton, Adam. 1994. "The Value of a Person." *Proceedings of the Aristotelian Society*, Supplementary Volume 68: 167–198.

Buchanan, Allen. 1995. "Equal Opportunity and Genetic Intervention." *Social Philosophy & Policy* 12: 105–135.

———. 1996. "Choosing Who Will Be Disabled: Genetic Intervention and the Morality of Inclusion." *Social Philosophy & Policy* 13: 18–46.

Buchanan, Allen, Brock, Dan, Daniels, Norman, and Wikler, Daniel. 2000. *From Chance to Choice: Genetics and Justice*. New York: Cambridge University Press.

Buckle, Stephen. 1988. "Arguing from Potential." *Bioethics* 2 (July): 222–238.

Buckle, Stephen, Dawson, Karen, and Singer, Peter. 1990. "The Syngamy Debate: When Precisely Does a Human Life Begin?" in Singer et al., eds., *Embryo Experimentation*. Cambridge: Cambridge University Press: 213–225.

Buckley, Rebecca, et al. 1999. "Hematopoietic Stem Cell Transplantation for the Treatment of Severe Combined Immunodeficiency." *New England Journal of Medicine* 341 (February 18): 508–513.

Bumol, Thomas, and Watanabe, August. 2001. "Genetic Information, Genomic Technologies, and the Future of Drug Discovery." *Journal of the American Medical Association* 285 (February 7): 551–555.

Cairns, John. 1997. *Matters of Life and Death: Perspectives on Public Health, Molecular Biology, Cancer and the Prospects for the Human Race*. Princeton, NJ: Princeton University Press.

Callahan, Daniel. 1987. *Setting Limits: Medical Goals in an Aging Society*. New York: Simon & Schuster.

————. 1990. *What Kind of Life? The Limits of Medical Progress.* New York: Simon & Schuster.

————. 1998. "Aging and the Allocation of Resources." *American Journal of Ethics and Medicine*: 4–9.

Cambridge Quarterly of Healthcare Ethics. 1998. Special issue on human cloning, Volume 7, Number 2.

Cassell, Eric. 1991. *The Nature of Suffering and the Goals of Medicine.* New York: Oxford University Press.

Cavazzano-Calvo, Marina, et al. 2000. "Gene Therapy for Human Combined Immunodeficiency (SCID)—X1 Disease." *Science* 288 (April 20): 669–672.

Childress, James, et al. 1997. "Cloning Human Beings: Responding to the National Bioethics Advisory Commission's Report." *Hastings Center Report* 27 (September-October): 9–22.

Churchland, Paul. 1984. *Matter and Consciousness.* Cambridge, MA: MIT Press.

Clark, William. 1995. *At War Within: The Double-Edged Sword of the Immune System.* New York: Oxford University Press.

————. 1997. *The New Healers: The Promise and Problems of Molecular Medicine in the Twenty-First Century.* New York: Oxford University Press.

Cloning Human Beings: Report and Recommendations of the National Bioethics Advisory Commission 1997. Rockville, MD.

Coleman, Jules, and Morris, Christopher (eds.). 1998. *Rational Commitment and Social Justice: Essays in Honor of Gregory Kavka.* Cambridge: Cambridge University Press.

Collins, Francis. 1999. "Medical and Societal Consequences of the Human Genome Project." *New England Journal of Medicine* 341 (January 7): 28–37.

Collins, Francis, and McKusick, Victor. 2001. "Implications of the Human Genome Project for Medical Science." *Journal of the American Medical Association* 285 (February 7): 540–544.

Cranor, Carl (ed.). 1994. *Are Genes Us? The Social Consequences of the New Genetics.* New Brunswick, NJ: Rutgers University Press.

Crobstein, C. 1964. "Cytodifferentiation and its Controls." *Science* 143: 643–650.

Cummings, Jeffrey. 1998. "Understanding Parkinson's Disease." *New England Journal of Medicine* 339 (October 8): 376–378.

Damasio, Antonio. 1994. *Descartes' Error: Reason, Emotion, and the Human Brain.* New York: Grosett/Putnam.

————. 1999. *The Feeling of What Happens: Body and Emotion in the Making of Consciousness.* New York: Harcourt Brace.

Daniels, Norman. 1985. *Just Health Care.* New York: Cambridge University Press.

———. 1988. *Am I My Parents' Keeper?* New York: Oxford University Press.

———. 1994. "The Genome Project, Individual Differences, and Just Health Care," in T. Murphy and M. Lappe, eds., *Justice and the Human Genome Project.* Berkeley: University of California Press: 110–132.

Delves, Peter, and Roitt, Ivan. 2000. "The Immune System," Parts I and II. *New England Journal of Medicine* 343 (July 6 and July 13): 37–49, 108–117.

Dobzhansky, Theodosius. 1975. "Evolution Is the Foundation of All Biology." *American Biology Teacher* 35: 125–129.

Draper, Heather, and Chadwick, Ruth. 1999. "Beware! Preimplantation Genetic Diagnosis May Solve Some Old Problems But It Also Raises New Ones." *Journal of Medical Ethics* 25: 114–120.

Duster, Troy. 1990. *Backdoor to Eugenics.* New York: Routledge.

Edelman, Gerald. 1987. *Neural Darwinism.* New York: Basic Books.

Edwards, R. G., and Purdy, J. 1981. *Human Conception in Vitro.* London: Academic Press.

Eisenberg, Leon. 1999. "Would Cloned Humans Really Be Like Sheep?" *New England Journal of Medicine* 340 (February 11): 471–475.

Eisenstadt v. Baird 1972. 405 U.S. 438, 453.

Elliot, Robert. 1997. "Genetic Therapy, Person-Regarding Reasons and the Determination of Identity." *Bioethics* 11 (April): 151–160.

Elliott, Carl. 1998. "The Tyranny of Happiness: Ethics and Cosmetic Psychopharmacology," in Parens, 177–188.

Evan, Gerald, and Littlewood, Trevor. 1998. "A Matter of Life and Cell Death." *Science* 281 (August 28): 1317–1325.

Ewald, P. W. 1994. *Evolution of Infectious Disease.* New York: Oxford University Press.

Eyre-Walker, Adam, and Keightley, Peter. 1999. "High Genomic Deleterious Mutation Rates in Hominids." *Nature* 397 (January 28): 344–347.

Feinberg, Joel. 1980. "The Child's Right to an Open Future," in W. Aiken and H. LaFollette, eds., *Whose Child?: Children's Rights, Parental Authority, and State Power.* Totowa, NJ: Rowman & Littlefield: 124–153.

———. 1984. *Harm to Others.* New York: Oxford University Press.

———. 1986. *Harm to Self.* New York: Oxford University Press.

———. 1992. "Wrongful Life and the Counterfactual Element in Harming," in Feinberg, *Freedom and Fulfillment.* Princeton, NJ: Princeton University Press: 3–36.

Feinberg, Todd. 2001. *Altered Egos: How the Brain Creates the Self.* New York: Oxford University Press.

Finch, Caleb. 1990. *Longevity, Senescence, and the Genome.* Chicago: University of Chicago Press.

Fischer, J. M., and Ravizza, M. 1998. *Responsibility and Control: A Theory of Moral Responsibility.* New York: Cambridge University Press.

Fletcher, Joseph. 1971. "Ethical Aspects of Genetic Controls." *New England Journal of Medicine* 285 (1971): 776–783.

———. 1974. *The Ethics of Genetic Control: Ending Reproductive Roulette.* Garden City, NY: Anchor Press.

Ford, Norman. 1988. *When Did I Begin?* Cambridge: Cambridge University Press.

Frankfurt, Harry. 1989. "Identification and Externality," in Frankfurt, *The Importance of What We Care About.* New York: Cambridge University Press: 58–68.

Fries, James. 1980. "Aging, Natural Death, and the Compression of Morbidity." *New England Journal of Medicine* 322 (July 17): 130–135.

———. 1983. "The Compression of Morbidity." *Milbank Memorial Fund Quarterly* 61: 347–419.

Galton, David, and Galton, Clare. 1998. "Francis Galton: And Eugenics Today." *Journal of Medical Ethics* 24: 99–105.

Gillam, Lynn. 1999. "Prenatal Diagnosis and Discrimination against the Disabled." *Journal of Medical Ethics* 25: 163–171.

Ginzberg, Eli. 1999. "Putting the Risk of Breast Cancer in Perspective." *New England Journal of Medicine* 340 (January 14): 141–146.

Glover, Jonathan. 1977. *Causing Death and Saving Lives.* Harmondsworth: Penguin.

———. 1984. *What Sort of People Should There Be?* Harmondsworth: Penguin.

———. 1992. "Future People, Disability, and Screening," in P. Laslett and J. Fishkin, eds., *Justice Between Age Groups and Generations.* New Haven, CT: Yale University Press: 127–143.

Gordon, Jon. 1999. "Genetic Enhancement in Humans." *Science* 283 (March 26): 2023–2024.

Govier, Trudy. 1982. "What Is Wrong with Slippery Slope Arguments?" *Canadian Journal of Philosophy* 12 (June): 303–316.

Grady, Denise. 2000. "Baby Conceived to Provide Cell Transplant for His Dying Sister." *New York Times,* October 4.

Greaves, Mel. 2000. *Cancer: The Evolutionary Legacy.* New York: Oxford University Press.

Grossman, M., et al. 1994. "Successful Ex Vivo Gene Therapy Directed to Liver in a Patient with Familial Hypercholesterolemia." *Nature Genetics* 6: 335–341.

Hanser, Matthew. 1990. "Harming Future People." *Philosophy & Public Affairs* 19 (Winter): 47–70.

Hardin, Garrett. 1995. *Living Within Limits*. New York: Oxford University Press.

Hardy, G. H. 1940. *Ramanujan: Twelve Letters on Subjects Suggested by His Life and Work*. Cambridge: Cambridge University Press.

Harris, John. 1997. "Goodbye Dolly? The Ethics of Human Cloning." *Journal of Medical Ethics* 23: 323–329

———. 1998. "Cloning and Human Dignity." *Cambridge Quarterly of Healthcare Ethics* 7 (Spring): 163–167.

———. 1998. *Clones, Genes, and Immortality*. Oxford: Oxford University Press.

Hart, Ronald, et al. 1997. "Born Again?" *The Sciences* 37: 47–51.

Hartman, Lynn, et al. 1999. "Efficacy of Bilateral Prophylactic Mastectomy in Women with a Family History of Breast Cancer." *New England Journal of Medicine* 340 (January 14): 77–84.

Hatcher, M. B., et al. 2001. "The Psychosocial Impact of Bilateral Prophylactic Mastectomy: Prospective Study Using Questionnaires and Semistructured Interviews." *British Medical Journal* 322 (January 13): 76–79.

Hayflick, Leonard. 1977. "The Cellular Basis for Biological Aging," in Hayflick and C. Finch, eds., *Handbook of the Biology of Aging*. New York: Van Nostrand: 159–186.

Heller, Jan Christian. 1996. *Human Genomic Research and the Challenge of Contingent Future Persons*. New York: Fordham University Press.

Heller, J., and Fotion, N., eds. 1997. *Contingent Future Persons: On the Ethics of Deciding Who Will Live, or Not, in the Future*. Boston: Kluwer.

Heyd, David. 1992. *Genethics: Moral Issues in the Creation of People*. Berkeley: University of California Press.

Hick, John. 1993. "A Possible Conception of Life After Death," in Hick, *Disputed Questions in Theology and the Philosophy of Religion*. New Haven, CT: Yale University Press: 183–196.

Holm, Soren. 1998. "A Life in the Shadow: One Reason Why We Should Not Clone Humans." *Cambridge Quarterly of Healthcare Ethics* 7 (Spring): 160–162.

Holtzman, Neil, and Marteau, Theresa. 2000. "Will Genetics Revolutionize Medicine?" *New England Journal of Medicine* 343 (July 13): 141–144.

———. 2000. "The Authors Reply." *New England Journal of Medicine* 343 (November 16): 1498.

Hubbard, Ruth, and Wald, Elijah. 1997. *Exploding the Gene Myth.* Boston: Beacon Press.

Hurka, Thomas. 1983. "Value and Population Size." *Ethics* 93 (April): 134–147.

———. 1993. *Perfectionism.* Oxford: Oxford University Press.

Jiang, Xu-Rong, et al. 1999. "Telomerase Expression in Human Somatic Cells Does Not Induce Changes Associated with a Transformed Phenotype." *Nature Genetics* 21 (January): 111–114.

Johnston, Mark. 1987. "Human Beings." *Journal of Philosophy* 84 (February): 59–83.

Jonas, Hans. 1992. "The Burden and Blessing of Morality." *Hastings Center Report* 22 (January-February): 34–40.

Juengst, Eric. 1997. "Can Enhancement Be Distinguished from Prevention in Genetic Medicine?" *Journal of Medicine and Philosophy* 22 (April): 125–142.

———. 1998. "What Does Enhancement Mean?" in Parens, 27–47.

Kagan, Shelly. 1989. *The Limits of Morality.* Oxford: Clarendon Press.

Kaji, Eugene, and Leiden, Jeffrey. 2001. "Gene and Stem Cell Therapies." *Journal of the American Medical Association* 285 (February 7): 545–550.

Kamm, F. M. 1992. "Non-Consequentialism, the Person as an End-in-Itself, and the Significance of Status." *Philosophy & Public Affairs* 21 (Fall): 354–389.

———. 1993. *Morality, Mortality: Volume I: Death and Whom to Save from It.* Oxford: Oxford University Press.

———. 1996. *Morality, Mortality: Volume II: Rights, Duties, and Status.* Oxford: Oxford University Press.

Kant, Immanuel. 1990. *Foundations of the Metaphysics of Morals* (1785), Second Edition, trans. L. W. Beck. New York: Macmillan.

Kass, Leon. 1985. *Toward a More Natural Science: Biology and Human Affairs.* New York: Free Press.

———. 1997. "The Wisdom of Repugnance: Why We Should Ban the Cloning of Humans." *The New Republic,* June 2: 17–26.

Kavka, Gregory. 1982. "The Paradox of Future Individuals." *Philosophy & Public Affairs* 11 (Spring): 93–112

———. 1994. "Upside Risks: Social Consequences of Beneficial Biotechnology," in Cranor, 155–179.

Keller, Gordon, and Snodgrass, Ralph. 1999. "Human Embryonic Stem Cells: The Future Is Now." *Nature Medicine* 5 (February): 151–152.

Kevles, Daniel. 1985. *In the Name of Eugenics: Genetics and the Uses of Human Heredity.* New York: Knopf.

Kevles, Daniel, and Hood, Leroy (eds.). 1992. *The Code of Codes: Scientific and Social Issues in the Human Genome Project*. Cambridge, MA: Harvard University Press.

Kitcher, Philip. 1996. *The Lives to Come: The Genetic Revolution and Human Possibilities*. New York: Simon & Schuster.

————. 1997. "Whose Self Is It, Anyway?" *The Sciences* 37, 58–62.

Kuhse, Helga. 1987. *The Sanctity-of-Life Doctrine in Medicine: A Critique*. Oxford: Clarendon Press.

Kuhse, Helga, and Singer, Peter. 1990. "Individuals, Humans, and Persons: The Issue of Moral Status," in Buckle, et al., eds., *Embryo Experimentation*: 80–100.

Lang, A. E., and Lozano, A. M. 1998. "Parkinson's Disease." *New England Journal of Medicine* 339 (October 8): 1044–1050.

Lappe, Marc. 1997. *The Tao of Immunology*. New York: Plenum Press.

Laslett, Peter, and Fishkin, James (eds.). 1992. *Justice Between Age Groups and Generations*. New Haven, CT: Yale University Press.

Lebacqz, Karen, and Geron Ethics Advisory Board. 1999. "Research with Human Embryonic Stem Cells: Ethical Considerations." *Hastings Center Report* 29: 31–36.

Lederberg, Joshua. 1963. "Molecular Biology, Eugenics, and Euphenics." *Nature* 198: 428–429.

Lewis, David. 1976. "Survival and Identity," in A. O. Rorty, ed., *The Identities of Persons*. Berkeley: University of California Press: 17–40.

Lewontin, Richard. 1999. *The Triple Helix: Gene, Organism, and Environment*. Cambridge, MA: Harvard University Press.

Lipkin, R. 1991. "The Quest to Break the Human Genetic Code." *Insight*, December-January: 46–48.

Lisman, John, and Fallon, Justin. 1999. "What Maintains Memories?" *Science* 283 (January 15): 339–340.

Lockwood, Michael. 1985. "When Does a Life Begin?" in Lockwood, ed., *Moral Dilemmas in Modern Medicine*. Oxford: Oxford University Press: 9–31.

McGinn, Colin. 1999. *The Mysterious Flame: Conscious Minds in a Material World*. New York: Basic Books.

McKerlie, Dennis. 1992. "Equality Between Age Groups." *Philosophy & Public Affairs* 21 (Summer): 273–295.

McKusick, Victor. 1992. *Mendelian Inheritance in Man: Catalogue of Autosomal Dominant, Autosomal Recessive, and X-Linked Phenotypes*, Tenth Edition. Baltimore: Johns Hopkins University Press.

McMahan, Jeff. 1996. "Cognitive Disability, Misfortune, and Justice." *Philosophy & Public Affairs* 25 (Winter): 3–35.

————. 1998. "Wrongful Life: Paradoxes in the Morality of Causing People to Exist," in Coleman and Morris, 208–247.

————. 2001. *The Ethics of Killing*. Oxford: Oxford University Press.

Maguire, G. Q., and McGee, Ellen. 1999. "Implantable Brain Chips? Time for Debate." *Hastings Center Report* 29 (January-February): 7–13.

Marshall, Eliot. 2000. "Gene Therapy on Trial." *Science* 288 (May 12): 951–957.

Marteau, Theresa, and Richards, Martin (eds.). 1996. *The Troubled Helix: Social and Psychological Implications of the New Human Genetics*. Cambridge: Cambridge University Press.

Martindale, Diane. 2001. "A Pink Slip in the Genes." *Scientific American*, January: 19–20.

Medawar, P. B. and Medawar, J. S. 1977. *The Life Science: Current Ideas of Biology*. New York: Harper & Row.

Medzhitov, Ruslan, and Janeway, Charles. 2000. "Innate Immunity." *New England Journal of Medicine* 343 (August 3): 338–344.

Mehlman, Maxwell, and Botkin, Jeffrey. 1998. *Access to the Genome*. Washington, DC: Georgetown University Press.

Meijers-Heijboer, Hanna, et al. 2001. "Breast Cancer After Prophylactic Bilateral Mastectomy in Women with a BRCA1 or BRCA2 Mutation." *New England Journal of Medicine* 345 (July 19): 159–164.

Michod, Richard. 1997. "What Good Is Sex?" *The Sciences* 37: 42–46.

Mirsky, Steve, and Rennie, John. 1997. "What Cloning Means for Gene Therapy." *Scientific American*, June: 122–123.

Mitchell, J., et al. 1993. "What Young People Think and Do When the Option for Cystic Fibrosis Carrier Testing Is Available." *Journal of Medical Genetics* 30: 538–542.

Mooney, David, and Mikos, Antonios. 1999. "Growing New Organs." *Scientific American*, April: 60–65.

Morales, Carmen, et al. 1999. "Absence of Cancer-Associated Changes in Human Fibroblasts Immortalized with Telomerase." *Nature Genetics* 21 (January): 115–118.

Mortensen, P. B., et al. 1999. "Effects of Family History and Place and Season of Birth on the Risk of Schizophrenia." *New England Journal of Medicine* 340 (February 25): 603–608.

Murphy, T., and Lappe, M. (eds.). 1994. *Justice and the Human Genome Project*. Berkeley: University of California Press.

Nagel, Thomas. 1979. "Death," in *Mortal Questions*. Cambridge: Cambridge University Press: 1–10.

―――. 1979. "What Is It Like to Be a Bat?" in *Mortal Questions*, 165–180.

―――. 1985. *The View from Nowhere*. New York: Oxford University Press.

―――. 1991. *Equality and Partiality*. Oxford: Oxford University Press.

Narveson, Jan. 1967. "Utilitarianism and New Generations." *Mind* 76: 67–72.

―――. 1973. "Moral Problems of Population." *The Monist* 57: 62–86.

Needham, J. A. 1959. *A History of Embryology*. New York: Abelard-Schuman.

Neel, James. 1997. "Looking Ahead: Some Genetic Issues of the Future." *Perspectives in Biology and Medicine* 40: 328–347.

Nesse, Randolph, and Williams, George. 1994. *Why We Get Sick: The New Science of Darwinian Medicine*. New York: Times Books.

―――. 1998. "Evolution and the Origins of Disease." *Scientific American*, November: 86–93.

Nussbaum, Martha, and Sunstein, Cass, eds. 1999. *Clones and Clones: Facts and Fantasies About Human Cloning*. New York: W. W. Norton.

Oderberg, David. 1996. "Coincidence Under a Sortal." *Philosophical Review* 105: 145–171.

―――. 1997. "Modal Properties, Moral Status, and Identity." *Philosophy & Public Affairs* 25 (Fall): 259–298.

Olshansky, S. Jay, and Carnes, Bruce. 2001. *The Quest for Immortality: Science at the Frontiers of Aging*. New York: Norton.

Olson, Eric. 1997. *The Human Animal: Personal Identity Without Psychology*. Oxford: Clarendon Press.

Parens, Eric (ed.). 1998. *Enhancing Human Traits: Ethical and Social Implications*. Washington, DC: Georgetown University Press.

Parfit, Derek. 1976. "Lewis, Perry, and What Matters," in Rorty, 91–107.

―――. 1984. *Reasons and Persons*. Oxford: Clarendon Press.

―――. 1986. "Comments." *Ethics* 96 (July): 832–872.

―――. 1995. "Equality or Priority?" *Lindley Lecture*. Lawrence, KS: University of Kansas Press.

Pence, Gerald. 1998. *Who's Afraid of Human Cloning?* Lanham, MD: Rowman & Littlefield.

Persson, Ingmar. 1995. "Genetic Therapy, Identity, and the Person-Regarding Reasons." *Bioethics* 9 (January): 16–31.

Pier, Gerald, et al., 1998. "Salmonella Typhi Uses CFTR to Enter Intestinal Epithelial Cells." *Nature* 393 (May 20): 79–82.

Promislow, D. 1998. "Longevity and the Barren Aristocrat." *Nature* 396 (December 24–31): 719–720.

—————. 1979. "Equality," in *Moral Questions*, 106–127.

Raikka, Juha. 1998. "Freedom and the Right (Not) to Know." *Bioethics* 12 (January): 49–63.

Rawls, John. 1971. *A Theory of Justice*. Cambridge, MA: Harvard Belknap Press.

—————. 1982. "Social Unity and Primary Goods," in Sen and Williams, 159–186.

Resnick, David. 1994. "Debunking the Slippery Slope Argument About Human Germ-Line Gene Therapy." *Journal of Medicine and Philosophy* 19 (February): 23–40.

Rhodes, Rosamond. 1998. "Genetic Links, Family Ties, and Social Bonds: Rights and Responsibility in the Face of Genetic Knowledge." *Journal of Medicine and Philosophy* 23 (February): 10–30.

Roitt, Ian. 1991. *Essential Immunology*, Seventh Edition. Oxford: Blackwell Scientific Publications.

Rose, Michael. 1991. *Evolutionary Biology of Aging*. Oxford: Oxford University Press.

Rosenfeld, Melissa. 1997. "Human Artificial Chromosomes Get Real." *Nature Genetics* 15: 333–335.

Ross, Russell. 1999. "Atherosclerosis—An Inflammatory Disease." *New England Journal of Medicine* 340 (January 14): 115–123.

Roth, D. A., et al. 2001. "Nonviral Transfer of the Gene Encoding Coagulation Factor VIII in Patients with Severe Hemophilia A." *New England Journal of Medicine* 344 (July 7): 1735–1742.

Rothstein, Mark. 1997. *Genetic Secrets: Protecting Privacy and Confidentiality in the Genetic Era*. New Haven, CT: Yale University Press.

Rowland, Lewis, and Shneider, Neil. 2001. "Amyotrohic Lateral Sclerosis." *New England Journal of Medicine* 344 (May 31): 1688–1700.

Sabin, James, and Daniels, Norman. 1994. "Determining 'Medical Necessity' in Mental Health Practice." *Hastings Center Report* 24: 5–13.

Scanlon, T. M. 1982. "Contractualism and Utilitarianism," in Sen and Williams, 103-128.

—————. 1998. *What We Owe to Each Other*. Cambridge, MA: Harvard University Press.

Scheffler, Samuel. 1992. *Human Morality*. New York: Oxford University Press.

Schuman, Michael. 1998. "Korean Experiment Fuels Cloning Debate: More Work Is Needed to Prove a Live Birth Is Possible." *Wall Street Journal*, December 21.

Science. 2001. Special Issue, "The Human Genome" 291 (February 16).

Sciences. 1997. Special Issue 37, "The Promise and Peril of Cloning."

Scientific American. 1999. Special Issue, "The Promise of Tissue Engineering," April.

Searle, John. 1992. *The Rediscovery of the Mind.* Cambridge, MA: MIT Press.

Sen, A., and Williams, B. (eds.). 1982. *Utilitarianism and Beyond.* Cambridge: Cambridge University Press.

Shapira, Amos. 1998. "Wrongful Life Lawsuits for Faulty Genetic Counseling: Should the Impaired Newborn Be Entitled to Sue?" *Journal of Medical Ethics* 24: 369–375.

Shiels, Paul, et al. 1999. "Analysis of Telomere Length in Cloned Sheep." *Nature* 339 (May 27): 316–317.

Shoemaker, Sydney. 1984. "Personal Identity: A Materialist's Account," in Shoemaker and Swinburne, *Personal Identity.* Oxford: Blackwell.

Silver, Lee. 1998. *Remaking Eden: How Genetic Engineering and Cloning Will Transform the American Family.* New York: Avon Books.

Solter, Davor. 1998. "Dolly *Is* a Clone—and No Longer Alone." *Nature* 394 (July 23): 315–316.

Southworth, Natalie. 2000. "Toronto Death Raises Questions About the Risks of Gene Therapy." *Globe and Mail,* March 6.

Steinbock, Bonnie. 1992. *Life Before Birth.* New York: Oxford University Press.

Steinbock, B., and McClamrock, R. 1994. "When Is Birth Unfair to the Child?" *Hastings Center Report* 24 (November-December): 363–369.

Stewart, B. W. 1994. "Mechanisms of Apoptosis: Integration of Genetic, Biochemical, and Cellular Indicators." *Journal of the National Cancer Institute* 86: 1286-1296.

Stocker, Michael. 1997. "Parfit and the Time of Value," in J. Dancy, ed., *Reading Parfit.* Oxford: Blackwell: 54–70.

Stolberg, Sheryl. 2000. "Youth's Death Shakes New Field of Gene Experiments on Humans." *New York Times,* January 27.

———. 2000. "A Second Death Linked to Gene Therapy." *New York Times,* May 4.

Swift, Jonathan. 1941. *Gulliver's Travels* (1726). London: Penguin.

Swinburne, Richard. 1984. "Personal Identity: The Dualist Theory," in Shoemaker and Swinburne, *Personal Identity.* Oxford: Blackwell.

———. 1997. *Evolution of the Soul,* Revised Edition. Oxford: Clarendon Press.

Suzuki, David, and Knudtson, Peter. 1990. *Genethics: The Clash Between the New Genetics and Human Values.* Cambridge, MA: Harvard University Press.

Tang, Ya-Ping, et al. 1999. "Genetic Enhancement of Learning and Memory in Mice." *Nature* 401 (September 2): 63–69.

Temkin, Larry, 1993. *Inequality.* Oxford: Oxford University Press.

Ten, C. L. 1998. "The Use of Reproductive Technologies in Selecting the Sexual Orientation, Race, and the Sex of Children." *Bioethics* 12 (January): 45–48.

Terry, R. D., et al. (eds.). 1999. *Alzheimer Disease*. Baltimore: Lippincott, Williams, and Wilkins.

Thomson, Judith Jarvis. 1990. *The Realm of Rights* Cambridge, MA: Harvard University Press.

Tooley, Michael. 1983. *Abortion and Infanticide*. Oxford: Clarendon Press.

Trevathan. W. R., et al. (eds.). 1999. *Evolutionary Medicine*. New York: Oxford University Press.

Unger, Peter. 1990. *Identity, Consciousness, and Value*. New York: Oxford University Press.

United States Congress Office of Technology Assessment. 1994. *Mapping Our Genes: Implications for Health and Social Policy*. Washington, DC: National Academy Press.

Van den Berg, Wibren. 1991. "The Slippery Slope Argument." *Ethics* 102 (October): 42–65.

Van Inwagen, Peter. 1990. *Material Beings*. Ithaca, NY: Cornell University Press.

Wade, Nicholas. 1998. "Immortality, of a Sort, Beckons Biologists." *New York Times*, November 17.

Wadman, Meredith. 1999. "Gene Therapy Pushes On, Despite Doubts." *Nature* 397 (January 14): 94.

Wagner, John. 1997. "Gene Therapy Is Not Eugenics." *Nature Genetics* 15: 23–24.

Wakayama, T., et al. 1998. "Full-Term Development of Mice from Enucleated Oocytes Injected with Cumulus Cell Nuclei." *Nature* 394 (July 23): 396–374.

Walters, Leroy, and Palmer, Julie Gage. 1997. *The Ethics of Human Gene Therapy*. New York: Oxford University Press.

Warren, Mary Ann. 1998. *Moral Status*. Oxford: Clarendon Press.

Wasserman, David, and Wachbroit, Robert, eds. 2001. *Genetics and Criminal Behavior*. New York: Cambridge University Press.

Westendorp, R., and Kirkwood, T. 1998. "Human Longevity at the Cost of Reproductive Success." *Nature* 396 (December 24–31): 743–744.

White, Stephen. 1989. *The Unity of the Self*. Cambridge, MA: MIT Press.

Wiggins, David. 1980. *Sameness and Substance*. Oxford: Blackwell.

Williams, Bernard. 1973. "The Makropulos Case: Reflections of the Tedium of Immortality," in Williams, *Problems of the Self*. Cambridge: Cambridge University Press: 82–100.

———. 1985. *Ethics and the Limits of Philosophy*. Cambridge, MA: Harvard University Press.

———. 1995. "Which Slopes Are Slippery?" in Williams, *Making Sense of Humanity*. Cambridge: Cambridge University Press: 213–223.

Williams, George. 1957. "Pleiotropy, Natural Selection, and the Evolution of Senescence." *Evolution* 1: 398–411.

Wilmut, Ian, et al. 1997. "Viable Offspring Derived from Fetal and Adult Mammalian Cells." *Nature* 385 (February 27): 310–313.

Wolffe, Alan, and Matzke, Marjori. 1999. "Epigenetics: Regulation Through Repression." *Science* 286 (October 15): 481–486.

Woodward, James. 1986. "The Non-Identity Problem." *Ethics* 96 (July): 804–831.

World Health Organization. 1986. Statement on Health and Disease.

INDEX

VEGF. *See* Vascular endothelial growth
    factor
Venter, Craig, 12, 176(n6)
Viral vectors. *See under* Gene therapy

Wakayama, T., 120, 191(n8)
Walters, Leroy, 102–103, 188(n26)
Watson, James, 12
Well-being. *See* Quality of life

Wikler, Daniel, 3–4
Williams, Bernard, 112
Williams, George C., 3, 4, 14, 19, 152, 153,
    155
Wilmut, Ian, 117, 119, 120, 191(n8)
Worse-off priority principle, 76–84, 98–99,
    101, 102, 157–158, 173, 186(n48)

Zygotes, 26–27, 28, 32, 179(n32)